SpringerBriefs in Physics

For further volumes:
http://www.springer.com/series/8902

Riccardo Fantoni

The Janus Fluid

A Theoretical Perspective

 Springer

Riccardo Fantoni
Dipartimento di Scienze dei Materiali e Nanosistemi
Università Ca' Foscari Venezia
Venice
Italy

ISSN 2191-5423   ISSN 2191-5431 (electronic)
ISBN 978-3-319-00406-8   ISBN 978-3-319-00407-5 (eBook)
DOI 10.1007/978-3-319-00407-5
Springer Cham Heidelberg New York Dordrecht London

Library of Congress Control Number: 2013936966

Printed on acid-free paper

Springer is part of Springer Science+Business Media (www.springer.com)

Foreword

Recent advances in the ability of creating functionalized colloidal solutions, able to self-assemble in a controlled way, have injected new reasons of interest in an already active field of research. The interest in new experimental methods for synthesizing and characterizing such systems is quite obvious from the point of view of applications. Probably less obvious is the challenge the new materials represent for liquid state theory. Very localized and anisotropic interactions dominate the relevant model interactions. The complexity of the resulting aggregation states calls for approximate methods to allow fast semi-quantitative exploration of the relevant thermodynamic space. Unfortunately, most of the classical approximations of the theory of liquids have to be adapted to the new class of interactions and their quality has to be assessed almost from scratch.

Such difficulties of the theory do not depend only on peculiarities of model interactions. Much more relevant is the new physics brought about by such models in the liquid and in the vapor phases. In particular, the microscopic structure is strongly affected by self-assembly phenomena resulting in a huge variety of locally preferred structures in the liquid and clusters in the vapor. The interplay and the effects of such complex microscopic structures on the phase diagrams is dramatic and certainly deserves careful investigations.

This brief monograph provides a timely insight into such a theme. Although the book is intended to provide an introduction to the state of the art of the theoretical description of the fluid phases of the so-called Janus colloids using the Kern and Frenkel model, it is self-contained and may be fruitfully used as a compact introduction to the basic theory useful to describe a broader class of systems.

Trieste, February 16, 2013   Giorgio Pastore

Preface

Today we are experiencing an unprecedented development in the synthesis of new colloidal particles in the laboratory. Among these, the Janus particles stem out for their simplicity and peculiar clustering properties. Janus particles can be viewed as one kind of fundamental colloidal molecules, whose molecular counterpart can be surfactant or dipolar molecules. Progress to synthesize such particles started to appear around 2003–2005. From 2006 onward the number of publications on Janus particles increased almost exponentially. Within the research work done from 1989 till today 63 % covers the synthesis and characterization, 16 % the assembly properties, 8 % the applications, 7 % the theory and simulation, and 6 % are reviews. In this brief book we report some of the theoretical advances that have been recently made in understanding the Janus fluid.

The book is composed of two chapters. In the first chapter the Janus fluid is described in its properties (phase diagram, structure, and clustering) and in its statistical physics problem. The state-of-the-art in the theoretical treatment of this rich and interesting fluid is given with a bridge between new research results published in journal articles and a contextual literature review. Some details about the Monte Carlo Metropolis algorithm used to study the fluid are presented. We gave only few details on the integral equation theories and on the thermodynamic perturbation theories because a full account can be found in the book of Hansen and McDonald [1].

The second chapter is devoted to the properties of clustering and self-assembly occurring in the Janus fluid and follow closely the results of Refs. [2, 3] where a cluster theory was constructed to describe semi-quantitatively the micellization properties of the fluid. These properties of the Janus fluid make it interesting in biological and medical physics as one could use the clustered superstructures as drug carriers in the human body.

The book is written at a graduate student level of comprehension and should be of some use for the student who wishes to undergo research work in the field of soft matter and biophysics or for the soft matter and biophysics scientist as a point of reference on the Janus fluid.

I wish to thank F. Sciortino, G. Pastore, and A. Giacometti for giving permission for the reproduction of Figs. 1.3, 1.4, 1.5 and for helpful suggestions in the writing of the manuscript.

Trieste   Riccardo Fantoni

Contents

1   **What is a Janus Fluid?**   1
   1.1   Introduction.   1
   1.2   The Classical Statistical Physics Problem   5
   1.3   Experimental Methods   7
      1.3.1   Measurements on a Macroscopic Scale   7
      1.3.2   Measurements on a Microscopic Scale   7
   1.4   Numerical Simulations   10
      1.4.1   Molecular Dynamics   10
      1.4.2   Monte Carlo.   11
   1.5   Relationship Between the Structure
      and the Thermodynamics   14
   1.6   The Integral Equation Approach   15
   1.7   The Thermodynamic Perturbation Approach   16
   1.8   The Phase Diagram of a Janus Fluid   17
   1.9   The Structure of a Janus Fluid.   18
   1.10 Perspectives.   20

2   **Clustering and Micellization in a Janus Fluid**   21
   2.1   Introduction.   21
   2.2   The Kern and Frenkel Model   23
   2.3   Clustering Properties   24
   2.4   A Cluster Theory for Janus Particles   26
   2.5   Relationship Between the Configurational
      Partition Functions.   27
      2.5.1   The Inter-Cluster Configurational Partition Function   27
      2.5.2   The Intra-Cluster Configurational Partition Function   30
      2.5.3   Thermodynamic Quantities.   33
   2.6   Results   33
      2.6.1   Case $\Delta = \sigma/2$.   34
      2.6.2   Case $\Delta = \sigma/4$.   38
      2.6.3   Case $\Delta = 3\sigma/20$ 39
   2.7   Conclusions.   39

Appendix A: Connection with Wertheim Association Theory   43

Appendix B: The Excess Internal Energy per Particle
            of the Clusters.   45

References   47

Index   49

Chapter 1
What is a Janus Fluid?

Unus mundus sum

Abstract The recent development of new sophisticated synthesis laboratory techniques in the field of colloidal particles allowed the realization of the Janus fluid in the laboratory and his characterization. In parallel, recent Monte Carlo simulations on the Kern and Frenkel model of the Janus fluid have revealed that in the vapor phase, below the critical point, there is the formation of preferred inert clusters made up of a well defined number of particles: the micelles and the vesicles. This is responsible for a re-entrant gas branch of the gas-liquid coexistence curve of the phase diagram. Detailed account of these new developments and new findings are given in the chapter where the Janus fluid is introduced at a statistical physics theoretical level outlining the progresses made in the use of the most common statistical physics instruments to study such a complex fluid, like the numerical simulations, the integral equation approach, and the thermodynamic perturbation approach.

Keywords Janus fluid · Janus particles · Kern-Frenkel model · Monte Carlo simulation · Integral equation theory · Thermodynamic perturbation theory · Radial distribution function · Structure factor · Phase diagram · Micelle · Vesicle · Lamella.

1.1 Introduction

A Janus fluid is one made of Janus particles immersed in a solvent. A Janus particle like the Roman God Janus, depicted in Fig. 1.1, is one that has two faces with two different functionalities. Originally, the term Janus particle was coined by Casagrande et al. in 1988 [4] to describe spherical glass particles with one of the hemispheres hydrophilic and the other hydrophobic. In that work, the amphiphilic beads were synthesized by protecting one hemisphere with varnish and chemically treating the

R. Fantoni, *The Janus Fluid*, SpringerBriefs in Physics,
DOI: 10.1007/978-3-319-00407-5_1, © The Author(s) 2013

Fig. 1.1 A statue representing
Janus Bifrons in the Vatican
Museums

Fig. 1.1 A statue representing Janus Bifrons in the Vatican Museums

other hemisphere with a silane reagent. This method resulted in a particle with equal hydrophilic and hydrophobic areas [5].

The Janus particles are commonly found amongst the soft matter colloidal particles [6]. Today an unprecedented development in particle synthesis is generating a whole new set of colloidal particles, characterized by different sizes, shapes, patterns, particle patchiness, and functionalities [7]. The concept of "Janus particles" was first raised by de Gennes 20 years ago in his Nobel Prize lecture [8]. What de Gennes had in mind was that these amphiphilic Janus particles might behave like surfactant molecules to adsorb at the water-air interface, forming a mono-layer that de Gennes described as a "skin that can breathe" since small molecules would still be able to diffuse through the interstices between the Janus particles in the mono-layer. Janus particles are the analogue to the surfactant molecule, which has a hydrophilic head group and hydrophobic tail(s). When we think about Janus particles, we usually consider particles with half-half geometry (hemispherical coverage): the "Janus limit". However, this is only one special case for Janus particles. Janus particles can have all kinds of different geometries. Janus balance is a concept to describe this characteristic. Janus balance is developed in analogous to the hydrophilic-lipophilic balance (HLB) value of small surfactant molecules. It is possible to quantify Janus balance by considering the adsorption energy of an amphiphilic Janus particle at the water-oil interface [9]. Similar to the HLB value, which determines many behaviors of small surfactants, Janus balance is a critical parameter to characterize Janus particles. It is important to be able to control Janus balance (geometry) during fabrication, since half-half geometry may not be the best for certain applications.

Most of the synthetic methods [10] can be roughly categorized by two different schemes. One is top-down [11, 12] and the other is bottom-up [13, 14]. Top down means the particles are fabricated using externally controlled device or process, such as lithography, or printing. Bottom-up is to produce particles in a self-assembly manner, where particles are formed by the organization of smaller pieces automatically. Usually, top-down method is used to synthesize particles larger than half-micron

(500 nm). Bottom-up methods are more versatile in making smaller particles down to few nanometers. Janus particles can be produced in large quantities [15] and they generally obey to Boltzmann statistics.

Exciting applications are on the horizon from using Janus particles to form useful structures using the bottom-up approach of spontaneous self-assembly. In this spirit, clusters of Janus particles have been observed to form spontaneously. Some resemble the micelles formed by conventional soap molecules but others are unexpectedly anisotropic in shape. Under microscope, some interesting assembly structures were observed; the micelle shape and the elongated string shapes formed when individual micelles polymerize into superstructures. It is well appreciated that, with the right balance of solvophilic and solvophobic moieties, surfactants can spontaneously self-assemble into a large variety of structures, including micelles, vesicles, lamellae. In some Janus particles, under typical experimental conditions in a water environment, one of the two hemispheres is hydrophobic, while the other is charged, so that different particles tend to repel each other, hence forming isolated monomers. On the other hand, if repulsive forces are screened by the addition of a suitable salt, then clusters tend to form driven by hydrophobic interactions [16]. Like wheels that rotate in a clock, rotation of any particle in these structures forces nearby particles to rotate also, just to preserve the requirement that hydrophobic sides must face one another. When a nearest neighbor rotates, so must the next-nearest neighbor and so on down the line, causing signal transduction over distances much longer than the particle itself. These ideas may also in the future enable aspects of responsive, adaptive structure. For example, at the sites of diseased tissues in the human body, mobile Janus particles may be triggered to form immobile clusters and drugs could subsequently be released from them, comprising smart new drug carriers [17]. Anisotropic surface charge makeup is interesting for applications even if one is not concerned with self-assembly. If a particle is coated black on one side and white on the other, and if an electric field can make it flip, then you have got a switchable optical element that can be a pixel in an optical display. Switchable particles also enable the design of optical lenses whose shape and focal length adjust to environmental stimuli. Such tasks are especially easy to accomplish when the particles are immersed in a fluid; hence the name of the field, liquid optics. Moreover, properly designed Janus particles are able to swim thanks to their anisotropic chemical makeup. The first example of self-propulsion came from catalytic reactions in which gas, evolved from a chemical reaction on one side of the particles, acted as a jet. Other examples are found amongst Janus particles immersed in an electric or a magnetic field [18].

The extensive effort in the synthesis of patchy colloidal particles is driven by the attempt to gain control over the three-dimensional organization of self-assembled materials, characterized by crystalline [19], or even disordered, colloidal complex superstructures with desired properties. This self-assembly mechanism has recently attracted increasingly attention due to the great improvement in the chemical synthetization and functionalization of such colloidal particles, that allows a precise and reliable control on the aggregation process that was not possible until few years ago [20]. From a technological point of view, this is very attractive as it paves the way to a bottom-up design and engineering of nanomaterials alternative to conventional top-down techniques [21].

One popular choice of microscopical mathematical model describing the typical duality characteristic of the Janus fluid is the Kern-Frenkel model [22]. This model considers a fluid of rigid spheres having their surfaces partitioned in two hemispheres: One having a square-well character, that is attracting other similar hemispheres through a square-well interactions, thus mimicking the short-range hydrophobic interactions occurring in real Janus fluids. The other part of the surface, is assumed to have hard-sphere interactions with all other hemispheres, that is both like hard-spheres as well as square-well hemispheres. The hard-sphere hemisphere hence models the charged part in the limit of highly screened interactions that is required to have aggregation of the clusters. Other models have also been considered [23].

Although in this book only an even distributions between square-well and hard-spheres surface distribution will be considered (Janus limit), other choices of the coverage, that is the fraction of square-well surface with respect to the total one, have been studied within the Kern-Frenkel model [24]. In fact, one of the most attractive feature of the model stems from the fact that it smoothly interpolates between an isotropic hard-sphere fluid (zero coverage) and an equally isotropic square-well fluid (full coverage) [25, 26].

The thermodynamical and structural properties of the Janus fluid have been recently investigated within the framework of the Kern-Frenkel model using numerical simulations [24, 27], thus rationalizing the cluster formation mechanism characteristic of the experiments [16]. The fluid-fluid transition was found to display unconventional and particularly interesting phase diagram, with a re-entrant transition associated with the formation of a cluster phase at low temperatures and densities [24, 27]. While numerical evidence of this transition is quite convincing, a minimal theory including all necessary ingredients for onset of this anomalous behavior is still missing. Two previous attempts are however noteworthy. Reinhardt et al. [28] introduced a van der Waals theory for a suitable mixture of cluster and monomers that accounts for a re-entrant phase diagram, whereas Fantoni et al. [2, 3], developed a cluster theory explaining the appearance of some "magic numbers" in the cluster formation. The Kern-Frenkel model in the "sticky limit" of a square well infinitely deep and infinitesimally wide is amenable to analytical solutions using particular integral equation theories [29] but the solution is unable to explain the anomaly present in the phase diagram of the Janus fluid. The challenge of a fully analytical theory describing the anomaly still remains [30].

The classical statistical physics problem of the Janus fluid, presented in the next Section, can be studied in its exact properties (like the phase diagram, the structure, and the clustering) through Monte Carlo simulations which are described in Sect. 1.4.2. Approximate properties can be extracted [1] through the use of integral equations [25, 26], thermodynamic perturbation [31], and cluster theories [2, 3]. A brief account of the integral equation theories is presented in Sect. 1.6 and of the thermodynamic perturbation theories in Sect. 1.7. The phase diagram of a particular model of the Janus fluid is presented in Sect. 1.8. The structure is presented in Sect. 1.9. And the clustering properties [2, 3] are presented in full details in the Chap. 2.

1.2 The Classical Statistical Physics Problem

We can study the Janus fluid using classical statistical physics [1]. For example in the canonical ensemble the systems of Janus particles, contained in a recipient of volume V, are imagined to have been brought into thermal equilibrium with each other by immersing them into a heat bath of temperature T. The canonical equilibrium probability density for a system of N identical Janus particles with Hamiltonian $\mathcal{H}_N$ coordinates $\mathbf{r}^N \equiv \{\mathbf{r}_1, \ldots, \mathbf{r}_N\}$ and momenta $\mathbf{p}^N \equiv \{\mathbf{p}_1, \ldots, \mathbf{p}_N\}$ is

$$f^{(N)}(\mathbf{r}^N, \mathbf{p}^N) = \frac{1}{N!h^{3N}} \frac{\exp[-\beta \mathcal{H}_N(\mathbf{r}^N, \mathbf{p}^N)]}{Q_N(V, T)}, \tag{1.1}$$

where h is Planck's constant, $\beta = 1/k_B T$ with k_B Boltzmann's constant, the factor $N!$ takes care of the indistinguishability of the particles, and the normalizing factor Q_N is the canonical partition function, defined as

$$Q_N(V, T) = \frac{1}{N!h^{3N}} \int \exp[-\beta \mathcal{H}_N(\mathbf{r}^N, \mathbf{p}^N)] d\mathbf{r}^N d\mathbf{p}^N. \tag{1.2}$$

Given a physical observable $\mathcal{O}(\mathbf{r}^N, \mathbf{p}^N)$ we can then measure its thermal average as follows

$$\langle \mathcal{O} \rangle = \int \mathcal{O}(\mathbf{r}^N, \mathbf{p}^N) f^{(N)}(\mathbf{r}^N, \mathbf{p}^N) d\mathbf{r}^N d\mathbf{p}^N. \tag{1.3}$$

The link between statistical mechanics and thermodynamics is established via the relation $F = -k_B T \ln Q_N(V, T)$, where F is the Helmholtz free energy. If the Hamiltonian is separated into kinetic and potential terms $\mathcal{H}_N = \sum_{i=1}^N p_i^2/2m + V_N(\mathbf{r}^N)$, the integration over momenta can be carried out explicitly, yielding a factor $(2\pi m k_B T)^{1/2}$ for each degree of freedom. The partition function may then be written as

$$Q_N(V, T) = \frac{1}{N! \Lambda^{3N}} Z_N(V, T), \tag{1.4}$$

where $\Lambda = (2\pi \beta \hbar^2/m)^{1/2}$ is the de Broglie thermal wavelength and

$$Z_N(V, T) = \int \exp[-\beta V_N(\mathbf{r}^N)] d\mathbf{r}^N, \tag{1.5}$$

is the configurational partition function. In the case of a perfect gas $V_N = 0$ and $Z_N^{id} = V^N$ where V is the volume enclosing the system of particles. The free energy per particle, using Stirling approximation, is $\beta F^{id}/N = \ln(\rho \Lambda^3) - 1$ where

$\rho = N/V$ is the density of the gas. The partition function of the system of interacting particles is conveniently written as

$$Q_N(V, T) = Q_N^{id} \frac{Z_N(V, T)}{V^N}. \tag{1.6}$$

On taking the logarithm of both sides the free energy separates naturally into "ideal" and "excess" parts: $F = F^{id} + F^{ex}$, where the excess part is

$$F^{ex} = -k_B T \ln \frac{Z_N(V, T)}{V^N}. \tag{1.7}$$

Given the free energy one can determine all the thermodynamical quantities like the pressure $P = -(\partial F/\partial V)_T$, the entropy $S = -(\partial F/\partial T)_V$, the internal energy $U = (\partial(F/T)/\partial(1/T))_V$, and the chemical potential $\mu = F/N + P/\rho$ or by thermodynamic integration from $N(\partial\mu/\partial N)_{V,T} = (\partial P/\partial\rho)_{N,T}$.

For Janus particles there is one more complication that is the fact that in order to know a configuration of the particles one has to specify the spatial coordinates $\mathbf{q}^N \equiv \{\mathbf{q}_1, \ldots, \mathbf{q}_N\}$ of the particles and their orientations $\omega^N \equiv \{\omega_1, \ldots, \omega_N\}$ relative to the laboratory frame so that $\mathbf{r}^N \equiv \{\mathbf{q}^N, \omega^N\}$ and $\mathbf{r}_i \equiv \{\mathbf{q}_i, \omega_i\}$ for all i. We can generally assume to have a pairwise interaction amongst the particles so that in the thermodynamic limit in a translational invariant (homogeneous) and rotational invariant (isotropic) fluid due to the translational and rotational symmetry we can write

$$V_N(\mathbf{r}^N) = \sum_{i<j}^N v(\mathbf{r}_i, \mathbf{r}_j) = \sum_{i<j}^N v(q_{ij}, \omega_i, \omega_j), \tag{1.8}$$

with $\mathbf{q}_{ij} = \mathbf{q}_j - \mathbf{q}_i$ and v the effective pair-potential between the Janus particles taking care of the effect of the solvent.

Another important tool needed in the statistical physics description of the fluid are the n-particle densities which in the canonical ensemble are defined as

$$\rho_N^{(n)}(\mathbf{r}^n) = \frac{N!}{(N-n)!} \frac{\int \exp[-\beta \mathcal{H}_N(\mathbf{r}^N, \mathbf{p}^N)] \, d\mathbf{r}^{(N-n)} d\mathbf{p}^N}{Q_N(V, T)}$$
$$= \frac{N!}{(N-n)!} \frac{\int \exp[-\beta V_N(\mathbf{r}^N)] \, d\mathbf{r}^{(N-n)}}{Z_N(V, T)}. \tag{1.9}$$

These define the structure of the fluid as will become more clear in Sect. 1.3.2. Of particular importance due to the direct link with the experiments is the 2-particle density $\rho_N^{(2)}(\mathbf{r}, \mathbf{r}') = \rho_N^{(1)}(\mathbf{r})\rho_N^{(1)}(\mathbf{r}')g(\mathbf{r}, \mathbf{r}')$ where g is known as the *pair correlation function* and $h = g - 1$ as the *total correlation function*. The 2-particle density gives the probability to find one particle at $\mathbf{r}$ and another particle at $\mathbf{r}'$. In the thermodynamic limit in an homogeneous and isotropic fluid the 1-particle density reduces to the usual density.

1.3 Experimental Methods

The experimental methods used when studying a real fluid fall in two categories: experiments which measure macroscopic quantities and those which measure microscopic quantities. The macroscopic data can usually be measured to a higher accuracy than the microscopic data.

1.3.1 Measurements on a Macroscopic Scale

Typical macroscopic measurements are experiments done to measure the pressure P, density ρ, and temperature T of a fluid. Integration of these measurements yields other thermodynamic quantities such as the internal energy, the heat capacities, or the compressibilities.

Measurements on a macroscopic scale are often needed to measure the transport coefficients of a fluid such as the shear and bulk viscosity, the thermal conductivity, or the diffusion coefficient.

1.3.2 Measurements on a Microscopic Scale

The most important class of microscopic measurements are the radiation scattering experiments. Among these three are particularly valuable: neutrons, X-rays, and laser light scattering. We will now give a brief description of a scattering experiment to stress the connection between measured quantities (the cross section) and theoretical concepts (the structure factor). For the sake of notational simplicity we will work with spherically symmetric particles so that a configuration of the particles will only be determined by their spatial coordinates $\mathbf{r}^N = \mathbf{q}^N$.

A typical layout of a scattering experiment on a fluid is shown in Fig. 1.2. The incident particles are wave packets with average momenta $\langle \mathbf{p} \rangle = \hbar \mathbf{k}_0$ and average impact parameter $\langle \rho \rangle$. They are assumed to be uniformly distributed on the $z = z_0 \rightarrow -\infty$ plane for $\rho \lesssim \rho_{max}$. The range of the scattering potential $\mathscr{V}(\mathbf{r})$ is $r_0 \ll \rho_{max}$.

We want to calculate the differential cross section $d\sigma/d\Omega$ defined as

$$\left[\frac{d\sigma}{d\Omega}(\theta, \phi) \right] d\Omega$$
$$\equiv \frac{\text{number of particles scattered in } d\Omega/\text{second}}{\text{number of incident particles}/(\text{second} \times \text{area on the } z = z_0 \text{ plane})}. \quad (1.10)$$

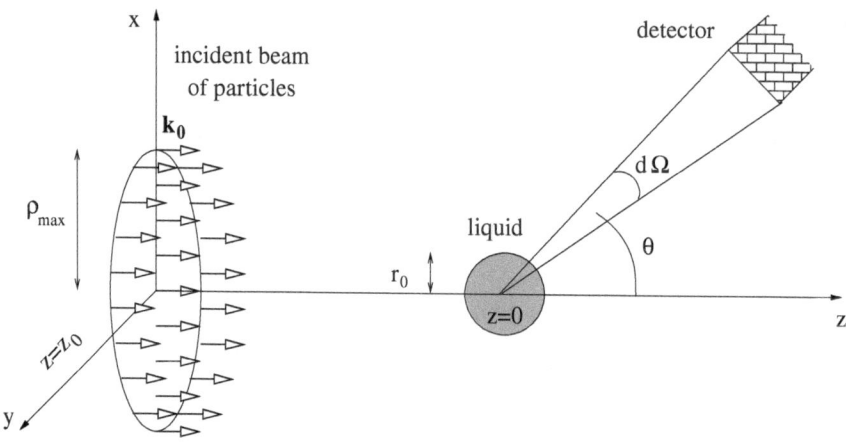

Fig. 1.2 Diagram showing a general scattering experiment for the measurement of the static structure factor. The incident beam of radiation is made up of particles with average momenta $\hbar k_0$, uniformly distributed on the $z = z_0 \to -\infty$ plane for $\rho = \sqrt{x^2 + y^2} \lesssim \rho_{max}$. The dimensions of the fluid r_0 are much smaller than ρ_{max}. The detector counts the number of scattered particles falling in the solid angle $d\Omega$ per second

Let us now assume, for simplicity, that the particles in the incident beam are neutrons.[1] The scattering of the neutron with the fluid occurs as a result of interactions with the atomic nuclei of the atoms of the fluid. These interactions are very short ranged, and the total scattering potential $\mathscr{V}(\mathbf{r})$ may therefore be approximated by a sum of delta function pseudopotentials of the form

$$\mathscr{V}(\mathbf{r}) = \frac{2\pi\hbar^2}{m} \sum_{i=1}^{N} b_i \delta(\mathbf{r} - \mathbf{r}_i), \tag{1.11}$$

where b_i is the scattering length of the ith nucleus. For most nuclei, b_i is positive, but it can also be negative and even complex; it varies both with isotopic species and with the spin state of the nucleus. Using the Born approximation one finds the following result for the differential cross section (see for example [32] Chap. 19)

$$\frac{d\sigma}{d\Omega} = \left\langle \sum_{i=1}^{N} \sum_{i=j}^{N} b_i b_j e^{-i\mathbf{k}\cdot(\mathbf{r}_i - \mathbf{r}_j)} \right\rangle, \tag{1.12}$$

where $\langle \ldots \rangle$ is the thermal average and $\mathbf{k} = \mathbf{k}_1 - \mathbf{k}_0$, with $\mathbf{k}_1 = k_0 \hat{r}$ the wave-vector of the particles collected by the detector. A more useful result is obtained taking a statistical average of the scattering lengths over both the isotopic species present in the sample and the spin states of the nuclei; this can be done independently of the

---

[1] Things are only slightly different for X-rays and light scattering. See later on in the text.

thermal averaging over the coordinates. We therefore introduce the notation

$$\langle b_i^2 \rangle = \langle b^2 \rangle,$$

$$\langle b_i b_j \rangle = \langle b_i \rangle \langle b_j \rangle = \langle b \rangle^2 = b_{coh}^2,$$

$$(\langle b^2 \rangle - \langle b \rangle^2) = b_{inc}^2,$$

where the subscript "coh" stands for coherent and "inc" for incoherent, and rewrite Eq. (1.12) as

$$\frac{d\sigma}{d\Omega} = N\langle b^2 \rangle + \langle b \rangle^2 \left\langle \sum_{i \neq j}^{N} e^{-i\mathbf{k} \cdot (\mathbf{r}_i - \mathbf{r}_j)} \right\rangle$$

$$= N(\langle b^2 \rangle - \langle b \rangle^2) + \langle b \rangle^2 \left\langle \left| \sum_{i=1}^{N} e^{-i\mathbf{k} \cdot \mathbf{r}_i} \right|^2 \right\rangle$$

$$= N b_{inc}^2 + N b_{coh}^2 S(\mathbf{k}). \qquad (1.13)$$

We then see that within the coherent contribution to the cross section appear the function $S(\mathbf{k})$ called the *static structure factor* of the fluid. It gives information on the structure of the fluid since for the homogeneous fluid its Fourier transform

$$\rho_N^{(2)}(\mathbf{r}, 0) \equiv \rho^2 g^{(2)}(\mathbf{r}) = \rho \int e^{i\mathbf{k} \cdot \mathbf{r}} [S(\mathbf{k}) - 1] \frac{d\mathbf{k}}{(2\pi)^3}, \qquad (1.14)$$

represents (see Eq. (1.9)) the probability density of finding a particle on the origin and another at $\mathbf{r}$. For a fluid that is also isotropic $g^{(2)}(r)$ is called the *radial distribution function* and $S(\mathbf{k}) = S(k)$.

A similar calculation can be made for the cross section of elastic scattering of X-rays. In this case only the coherent part gives a contribution and, since X-rays are scattered by the atomic electrons, the analog of b is the atomic form factor.

When the energy of the incident particles is comparable to the thermal energies of the atoms of the fluid, as for thermal neutrons, the scattering cannot be considered elastic any more. The cross section can therefore be measured as a function of energy transfer as well as momentum transfer. By this means it is possible to extract information on wave-number and frequency dependent fluctuations in fluids at wavelengths comparable with the spacing between particles (see [1] Chap. 7). Light scattering experiments yield similar results to thermal neutron scattering, but the accessible range of momentum transfer limits the method to the study of fluctuations of wavelengths of order 10^{-5}cm more appropriate for the Janus fluid.

The structure of a single Janus particle or cluster is usually accessible through the use of images from a scanning electron microscope (SEM) [33].

1.4 Numerical Simulations

Numerical simulations of classical fluids [34, 35], some times called computer experiments, can be of two types: the ones using the method of molecular dynamics [36]
and the ones using the Monte Carlo method of Metropolis [37].

These computer experiments give exact results for the particular model studied.
Since computers cannot deal with an Avogadro's number of particles the usefulness
of these methods rests in the fact that a model containing a relatively small number of
particles (several hundreds) is in most cases sufficiently large to simulate the behavior
of a macroscopic system when periodic boundary conditions [38] are employed.
Moreover with a computer experiment is possible to obtain information on quantities
of theoretical importance that are not readily measurable in the laboratory.

Molecular dynamics is especially valuable since it allows the study of time dependent phenomena. While to study the static properties of a system the Monte Carlo
method is often more suitable, primarily because the implementation of phase averages in any statistical ensemble is simpler than in Molecular dynamics.

1.4.1 Molecular Dynamics

In a typical molecular dynamic calculation a system of N particles is placed in
a cubical box of fixed volume with periodic boundary conditions. A set of initial velocities is assigned to each particle. The velocities are usually drawn from
a Maxwell-Boltzmann distribution appropriate to the temperature of interest and
selected in such a way as to have the net linear momentum initially equal to zero.

The trajectory of the particles are then calculated by integration of the classical
equations of motion

$$m_i \ddot{\mathbf{q}}_i = \mathbf{f}_i = -\nabla_i V_N(\mathbf{r}^N), \tag{1.15}$$

where m_i is the mass of particle i, $\mathbf{q}_i$ is its position, a dot denotes a total time
derivative, and V_N is the total potential energy already introduced in Eq. (1.8). The
dynamical states that the method generates represent a sample from a microcanonical
ensemble.

In the early stages of the calculation it is normal for the temperature to drift away
from the value at which it was originally set, and an occasional rescaling of the
particles velocities is therefore necessary. Once equilibrium is reached, the system
is allowed to evolve undisturbed, with both kinetic and potential energies fluctuating
around steady mean values.

The coordinates $\mathbf{r}^N$ and momenta $\mathbf{p}^N$ of the particles are stored for later analysis.
For example if $\mathscr{O}(\mathbf{r}^N, \mathbf{p}^N)$ is a function of the coordinates and momenta, and O is
the associated thermodynamic property, the simplest way to obtain O is through a
time average of $\mathscr{O}$ over the dynamical history of the system

$$O = \langle \mathscr{O} \rangle_t = \lim_{\tau \to \infty} \frac{1}{\tau} \int_0^\tau \mathscr{O}[\mathbf{r}^N(t), \mathbf{p}^N(t)] \, dt. \tag{1.16}$$

1.4.2 Monte Carlo

Apart from the choice of initial conditions, a molecular dynamics simulation is, in principle, entirely deterministic in nature. By contrast, as the name suggests, any Monte Carlo computation is essentially probabilistic.

The canonical ensemble average of any property $\mathcal{O}$ function of the particles coordinates can be written as

$$\langle \mathcal{O} \rangle = \frac{\int \mathcal{O}(\mathbf{r}^N) e^{-\beta V_N(\mathbf{r}^N)} \, d\mathbf{r}^N}{\int e^{-\beta V_N(\mathbf{r}^N)} \, d\mathbf{r}^N}. \tag{1.17}$$

The presence of multidimensional integrals rules out the possibility to use deterministic quadrature methods to calculate $\langle \mathcal{O} \rangle$. We could instead attempt to generate a large number of random configurations of particles $\{s_0, s_1, s_2, \ldots, s_P\}$ with $s_i = (\mathbf{r}^N)_i$ and evaluate $\langle \mathcal{O} \rangle$ as

$$\langle \mathcal{O} \rangle \simeq \frac{\sum_{m=1}^{P} \mathcal{O}(s_m) e^{-\beta V_N(s_m)}}{\sum_{m=1}^{P} e^{-\beta V_N(s_m)}}. \tag{1.18}$$

This crude approach is in practice very inefficient because a randomly constructed configuration is likely to have a very small Boltzmann factor.

It is then necessary to introduce importance sampling [39], i.e sample configurations in such a way that the regions of configuration space that make the largest contribution to the integrals in Eq. (1.17) are also the regions that are sampled most frequently. If $\pi(s_m)$ is the probability of choosing a configuration m, Eq. (1.18) must be replaced by

$$\langle \mathcal{O} \rangle \simeq \frac{\sum_{m=1}^{P} \mathcal{O}(s_m) e^{-\beta V_N(s_m)} / \pi(s_m)}{\sum_{m=1}^{P} e^{-\beta V_N(s_m)} / \pi(s_m)}. \tag{1.19}$$

If one can sample on the Boltzmann distribution itself

$$\pi(s_m) = \frac{e^{-\beta V_N(s_m)}}{\sum_{m=1}^{P} e^{-\beta V_N(s_m)}}, \tag{1.20}$$

(1.19) reduces to

$$\langle \mathcal{O} \rangle \simeq \frac{1}{P} \sum_{m=1}^{P} \mathcal{O}(s_m). \tag{1.21}$$

However, in the usual statistical mechanics calculations, the normalization denominator in (1.20) is not known and only relative probabilities of different configurations are easily accessible. The problem of finding a scheme for sampling configuration space according to a specific probability distribution is most easily formulated in terms of the theory of stochastic processes.

In a random walk (Markov chain) one changes the state of the system randomly according to a fixed transition rule $\mathscr{P}(s \rightarrow s')$, thus generating a random walk through state space $\{s_0, s_1, s_2, \ldots\}$. The definition of a Markov process is that the next step is chosen from a probability distribution that depends only on the "present" position. $\mathscr{P}(s \rightarrow s')$ is a probability distribution so it satisfies

$$\sum_{s'} \mathscr{P}(s \rightarrow s') = 1, \tag{1.22}$$

and

$$\mathscr{P}(s \rightarrow s') \geq 0. \tag{1.23}$$

The transition probability often satisfies the detailed balance property: the transition rate from s to s' equals the reverse rate

$$\pi(s)\mathscr{P}(s \rightarrow s') = \pi(s')\mathscr{P}(s' \rightarrow s). \tag{1.24}$$

If the pair $\pi(s)$, $\mathscr{P}(s \rightarrow s')$ satisfies the detailed balance and if $\mathscr{P}(s \rightarrow s')$ is ergodic,[2] then the random walk must eventually have π as its equilibrium asymptotic distribution. Detailed balance is one way of making sure that we sample π; it is a sufficient condition.

The Metropolis (rejection) method is a particular way of ensuring that the transition rules satisfy detailed balance. It does this by splitting the transition probability into an "a priori" sampling distribution $T(s \rightarrow s')$ (a probability distribution that we can directly sample) and an acceptance probability $A(s \rightarrow s')$ where $0 \leq A \leq 1$

$$\mathscr{P}(s \rightarrow s') = T(s \rightarrow s')A(s \rightarrow s'). \tag{1.25}$$

In the generalized Metropolis procedure [40], trial moves are accepted according to

$$A(s \rightarrow s') = \min[1, q(s \rightarrow s')], \tag{1.26}$$

where

$$q(s \rightarrow s') = \frac{\pi(s')T(s' \rightarrow s)}{\pi(s)T(s \rightarrow s')}. \tag{1.27}$$

It is easy to verify detailed balance and hence asymptotic convergence with this procedure by looking at the three cases: $s = s'$ (trivial), $q(s \rightarrow s') \leq 1$, and $q(s \rightarrow s') \geq 1$.

---

[2] Ergodicity is ensured if: (1) one can move from any state to any other state in a finite number of steps with a nonzero probability, (2) the transition probability is not periodic (always true if $\mathscr{P}(s \rightarrow s) > 0$), (3) the average return time to any state is finite. This is always true in a finite system (e.g. periodic boundary conditions).

This is the generalized Metropolis algorithm:

1. Decide what distribution to sample $[\pi(s)]$ and how to move from one state to another $T(s \rightarrow s')$.
2. Initialize the state, pick s_0.
3. To advance the state from s_n to s_{n+1}:

   - Sample s' from $T(s_n \rightarrow s')$.
   - Calculate the ratio

   $$q = \frac{\pi(s')T(s' \rightarrow s_n)}{\pi(s_n)T(s_n \rightarrow s')}. \tag{1.28}$$

   - Accept or reject: if $q > u_n$ where u_n is a uniformly distributed random number in $(0,1)$ set $s_{n+1} = s'$, otherwise set $s_{n+1} = s_n$.
4. Throw away the first κ states as being out of equilibrium (κ being the "warm-up" time).
5. Collect averages every so often and block them to get error bars.

Consider the sampling of the classical Boltzmann distribution $\exp(-\beta V_N(s))$. In the original Metropolis procedure [37], $T(s \rightarrow s')$ was chosen to be a constant distribution. This is the "classic" rule: a single particle position at a single "time" slice is displaced uniformly inside a cube and the cube side Δ is adjusted to achieve an efficient sampling of the configuration position space. Its orientation is chosen randomly on the surface of a sphere according for example to the Marsaglia algorithm (see Ref. [35] appendix G.4). Since T is a constant, it drops out of the acceptance formula. The acceptance is based on $q = \exp\{-\beta[V_N(s') - V_N(s)]\}$: moves that lower the potential energy are always accepted, moves that raise the potential energy are often accepted if the energy cost (relative to $1/\beta$) is small.

Some things to note about Metropolis:

- The acceptance ratio (number of successful moves/total number of trials) is a key quantity to keep track of and to quote. If it is very small one is doing a lot of work without moving through phase space, if it is close to 1 one could use larger steps and get faster convergence.
- One nice feature is that particles can be moved one at a time.
- The normalization of π is not needed, only ratios enter in.
- One can show that Metropolis acceptance formula is optimal among formulas of this kind which satisfy detailed balance (the average acceptance ratio is as large as possible).

A Monte Carlo simulation can be performed [41] at constant NVT (in the canonical ensemble, the one described in this section), at constant NPT (in the isobaric-isothermal ensemble), at constant μVT (in the grand-canonical ensemble). There is then the Gibbs ensemble Monte Carlo method which is used specifically to determine the gas-liquid coexistence curve of a given fluid.

1.5 Relationship Between the Structure and the Thermodynamics

The assumed absence of three-body forces between the particles (see Eq. (1.8)) allows to determine the thermodynamical quantities given the radial distribution function of the fluid as follows. For example the excess internal energy is

$$U^{ex} = \frac{1}{Z_N(V, T)} \int \exp[-\beta V_N(\mathbf{r}^N)] \left(\frac{1}{2} \sum_{i \neq j}^{N} v(\mathbf{r}_i, \mathbf{r}_j) \right) d\mathbf{r}^N. \qquad (1.29)$$

The double sum over i, j gives rise to $N(N-1)/2$ terms, each of which yields the same result after integration. The integral may therefore be rewritten as

$$
\begin{aligned}
U^{ex} &= \frac{N(N-1)}{2} \int v(\mathbf{r}_1, \mathbf{r}_2) \left(\frac{1}{Z_N(V, T)} \right. \\
&\quad \left. \times \int \exp[-\beta V_N(\mathbf{r}^N)] d\mathbf{r}_3 \cdots d\mathbf{r}_N \right) d\mathbf{r}_1 d\mathbf{r}_2 \\
&= \frac{\rho^2}{2} \int v(\mathbf{r}_1, \mathbf{r}_2) g^{(2)}(\mathbf{r}_1, \mathbf{r}_2) d\mathbf{r}_1 d\mathbf{r}_2 \\
&= 2\pi \rho N \int_0^{\infty} \langle v(\mathbf{r}_1, \mathbf{r}_2) g^{(2)}(\mathbf{r}_1, \mathbf{r}_2) \rangle_{\omega_1 \omega_2} q_{12}^2 \, dq_{12},
\end{aligned} \qquad (1.30)
$$

where we use an angular bracket with subscript $\omega_1 \cdots$ to denote an unweighted average over the angles $\omega_1 \cdots$. Thus

$$\langle \cdots \rangle_{\omega_1} \equiv \frac{1}{\omega} \int \cdots d\omega_1, \qquad (1.31)$$

where for Janus particles $\omega = 4\pi$.

The equation of state can also be expressed in terms of $g^{(2)}$. From the virial theorem [1] follows that

$$
\begin{aligned}
\frac{\beta P}{\rho} &= 1 - \frac{\beta}{3N} \left\langle \sum_{i=1}^{N} \mathbf{q}_i \cdot \nabla_i V_N(\mathbf{r}^N) \right\rangle \\
&= 1 - \frac{\beta}{3N Z_N(V, T)} \int \exp[-\beta V_N(\mathbf{r}^N)] \sum_{i=1}^{N} \mathbf{q}_i \cdot \nabla_i \left(\sum_{j \neq i, j=1}^{N} v(\mathbf{r}_i, \mathbf{r}_j) \right) d\mathbf{r}^N \\
&= 1 - \frac{\beta}{6N} \sum_{i \neq j}^{N} \frac{1}{Z_N(V, T)} \int \exp[-\beta V_N(\mathbf{r}^N)] \mathbf{q}_{ij} \cdot \nabla_{ij} v(\mathbf{r}_i, \mathbf{r}_j) d\mathbf{r}_N \\
&= 1 - \frac{2\pi \beta \rho}{3} \int_0^{\infty} \langle v'(\mathbf{r}_1, \mathbf{r}_2) g^{(2)}(\mathbf{r}_1, \mathbf{r}_2) \rangle_{\omega_1 \omega_2} q_{12}^3 \, dq_{12},
\end{aligned} \qquad (1.32)
$$

where the prime denotes differentiation with respect to q_{12} with ω_1, ω_2 held constant.

Irrespective of whether or not the potential energy is pairwise-additive, the isothermal compressibility $\chi_T \equiv -(\partial V/\partial P)_T/V$, is given by

$$\rho k_B T \chi_T = 1 + \rho \int \langle g^{(2)}(\mathbf{r}_1, \mathbf{r}_2) - 1 \rangle_{\omega_1 \omega_2} d\mathbf{q}_{12}. \tag{1.33}$$

It should be noted that there exist other routes to the equation of state other than the virial route described in Eq. (1.32). Like for example the compressibility route where one determines the pressure from thermodynamic integration of the isothermal compressibility or the energy route where one determines the pressure from thermodynamic integration of the internal energy per particle $u = U/N$ from the Maxwell relation $(\partial \beta P/\partial \beta)_N = \rho^2(\partial u/\partial \rho)_T$. Naturally in an exact calculations as the one carried out in a Monte Carlo integration the three routes will give the same results but if one uses approximate theories like integral equations then there may be inconsistencies between the various routes.

1.6 The Integral Equation Approach

The structure of the fluid can be extracted approximately through the use of integral equation theories [1]. These are based on the so called Ornstein-Zernike (OZ) equation and a closure. The OZ equation can be written in terms of the indirect correlation function $\gamma(\mathbf{r}_1, \mathbf{r}_2) = h(\mathbf{r}_1, \mathbf{r}_2) - c(\mathbf{r}_1, \mathbf{r}_2)$ and the direct correlation function c as follows

$$\gamma(\mathbf{r}_1, \mathbf{r}_2) = \rho \int d\mathbf{q}_3 \langle [\gamma(\mathbf{r}_1, \mathbf{r}_3) + c(\mathbf{r}_1, \mathbf{r}_3)]c(\mathbf{r}_3, \mathbf{r}_2) \rangle_{\omega_3}. \tag{1.34}$$

The closure in its general form is written as

$$c(\mathbf{r}_1, \mathbf{r}_2) = \exp[-\beta v(\mathbf{r}_1, \mathbf{r}_2) + \gamma(\mathbf{r}_1, \mathbf{r}_2) + B(\mathbf{r}_1, \mathbf{r}_2)] - 1 - \gamma(\mathbf{r}_1, \mathbf{r}_2). \tag{1.35}$$

The Bridge function B is known only as an infinite power series in density whose coefficients cannot be readily calculated. All practical closures approximate B in some way. Then a solution of Eqs. (1.34) and (1.35) together gives the total correlation function or the radial distribution function.

In general it is necessary to solve the integral equation through numerical means. The one used more frequently is the Newton-Raphson, which allows to find the common roots of the two non-linear operators $F_1(\gamma, c) = \hat{\gamma} - \rho \hat{c}(\hat{\gamma} + \hat{c})$ and $F_2(\gamma, c) = closure$. Where the hat over the correlation functions denotes a Fourier transform and once the inverse space or momentum space is discretized, $\hat{\gamma}$ and $\hat{c}$ become vectors of a very high number of components. And we are assuming for simplicity that we are dealing with spherically symmetric particles so that a configuration of the particles will only be determined by their spatial coordinates $\mathbf{r}^N = \mathbf{q}^N$.

In Ref. [25] the reference hypernetted-chain closure [1] is used to solve for the structure of the particularly simple Kern-Frenkel model of the Janus fluid. The one described by the simple pair-potential of Eq. (2.1). In this case the broken spherical symmetry carried through by the angular dependence of the pair potential can be dealt with expanding into spherical harmonics any correlation function in its axial frame and rewriting the Ornstein-Zernike equation in momentum space in matrix form as explained in Appendix A of Ref. [25].

1.7 The Thermodynamic Perturbation Approach

Given a model fluid with a total potential energy written in the following form [1]

$$V_\gamma(\mathbf{r}^N) = V_0(\mathbf{r}^N) + \gamma V_I(\mathbf{r}^N), \tag{1.36}$$

where $V_0(\mathbf{r}^N) = \sum_{i<j} v_0(\mathbf{r}_i, \mathbf{r}_j)$ is the unperturbed part, $V_I(\mathbf{r}^N) = \sum_{i<j} v_I(\mathbf{r}_i, \mathbf{r}_j)$ is the perturbation part, and $0 \leq \gamma \leq 1$ is used as perturbation parameter, one can determine the second order perturbation result for the Helmholtz free energy as

$$F_\gamma = F_0 + \gamma \left(\frac{\partial F_\gamma}{\partial \gamma}\right)_{\gamma=0} + \frac{1}{2!}\gamma^2 \left(\frac{\partial^2 F_\gamma}{\partial \gamma^2}\right)_{\gamma=0} + \ldots \tag{1.37}$$

In Ref. [31] the second order thermodynamic perturbation theory is applied to the simple Kern-Frenkel model of the Janus fluid where the pair-potential of Eq. (2.1) is rewritten in the following way

$$v(\mathbf{r}_1, \mathbf{r}_2) = v_0(q_{12}) + v_I(\mathbf{r}_1, \mathbf{r}_2), \tag{1.38}$$

where the first term is the hard-sphere contribution

$$v_0(q) = \begin{cases} \infty & 0 < r < \sigma \\ 0 & r > \sigma \end{cases}, \tag{1.39}$$

and the second term

$$v_I(\mathbf{r}_1, \mathbf{r}_2) = v_{SW}(q_{12})\Psi(1, 2), \tag{1.40}$$

is the orientation-dependent attractive part which can be factorized into an isotropic square-well tail of width $\Delta = (\lambda - 1)/\sigma$

$$v_{SW} = \begin{cases} -\varepsilon & \sigma < r < \lambda\sigma \\ 0 & r > \lambda\sigma \end{cases}, \tag{1.41}$$

multiplied by the angular dependent factor Ψ defined in Eq. 2.3. In Appendices A and B of Ref. [31] the extension of the Barker-Henderson thermodynamic perturbation scheme [1] to the angular dependent potential is reported. Once the free energy is known one can determine for example the gas-liquid coexistence curve in the phase diagram by equating the chemical potentials and the pressures of the two phases at a given temperature and compare with the one from the exact Monte Carlo simulations described in the next section.

1.8 The Phase Diagram of a Janus Fluid

Monte Carlo simulations in the Gibbs ensemble [42–44] have now been successfully used for several years to study first-order phase transitions in fluids. For temperatures far below the critical point, satisfactory results for the phase coexistence densities can be obtained. Near the critical point, however, finite size effects become significant, and finite size scaling [45] has proven to be important in the study of fluids in the continuum [46]. The Gibbs ensemble Monte Carlo method of Panagiotopoulos is now widely adopted as a standard method for calculating phase equilibria from molecular simulations. According to this method, the simulation is performed in two boxes containing the coexisting phases. Equilibration in each phase is guaranteed by moving particles. Equality of pressures is satisfied in a statistical sense by expanding the volume of one of the boxes and contracting the volume of the other. Chemical potentials are equalized by transferring particles from one box to the other.

Sciortino et al. [27] determined through Gibbs ensemble Monte Carlo [41–44, 47] simulations the gas-liquid coexistence curve, the *binodal*, for the Kern-Frenkel model of the Janus fluid. They found the result shown in Fig. 1.3.

The same model fluid has also been studied through the use of thermodynamic perturbation theory in Ref. [31]. The theory is unable to explain the anomalous re-entrant behavior of the gas branch of the coexistence curve but explains the shift at higher temperatures of the critical point as the attractive patch coverage of the particle surface increases above 50 %.

As we can see from the Fig. 1.3 the gas branch of the binodal tends to shift at high densities at low temperature and this re-entrant feature which is a peculiarity of the fluid at the Janus limit [24] is due to the formation, in the vapor phase below the critical point of the fluid, of inert clusters: the micelles (made up of around ten particles, see Fig. 1.4) and the vesicles (made up of around forty particles, see Fig. 1.4). This clustering phenomena will be described in more detail in the Chap. 2. The cluster tend to be inert because they are formed by an ensemble of particles which tend to display themselves with the inert hemisphere on the outer region of the cluster in order to maximize favorable contacts. So the interaction between two clusters tends to be very small. And this pushes the vapor phase to higher densities. At low temperatures and high densities the fluid enters the lamellar solid phase.

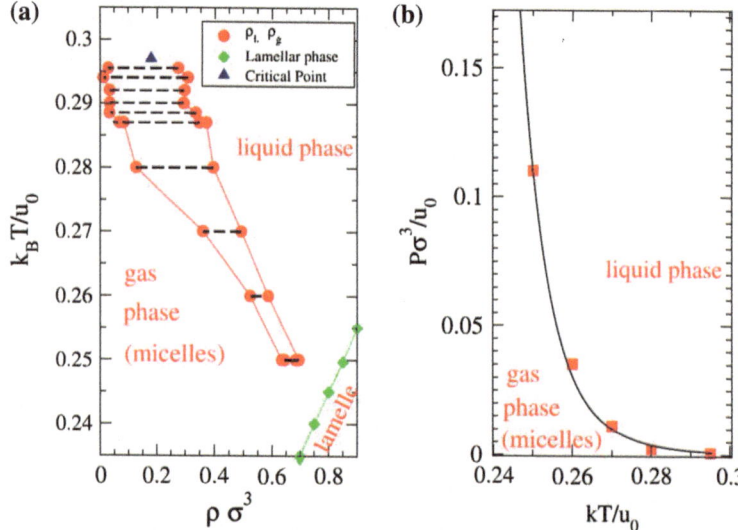

Fig. 1.3 Phase diagram of the Janus model for $\Delta/\sigma = 1/2$ in the $T - \rho$ (**a**) and $P - T$ (**b**) planes. Symbols are simulation results, lines are a guide to the eyes. The figure is from Ref. [27] and their u_0 is our ε. ρ_l and ρ_g are the coexisting density of the liquid and of the gas phase respectively

Fig. 1.4 Typical clusters formed in the vapor phase of the Kern-Frenkel Janus fluid: **a** the micelle and **b** the vesicle. The *red* particle hemisphere is inert and the *green* one is active, attractive. The Figure is from Ref. [27]

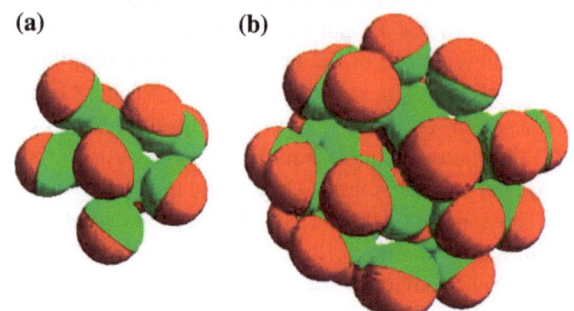

1.9 The Structure of a Janus Fluid

It is particularly relevant to look at the structure factor in this system both in the gas phase at low T, when micelles and vesicles become prominent, as well as in the liquid phase. Sciortino et al. [24] for the Kern-Frenkel model of the Janus fluid with the pair-potential of Eq. (2.1) obtained through canonical Monte Carlo simulations the static structure factors described in Fig. (1.5).

The same model fluid has also been studied through the use of integral equation theory in Refs. [25, 26] and of thermodynamic perturbation theory in Ref. [31].

The first panel of Fig. 1.5 shows the evolution of the structure factor on increasing density at $k_B T/\sigma = 0.25$, across the gas-liquid transition. At this low temperature

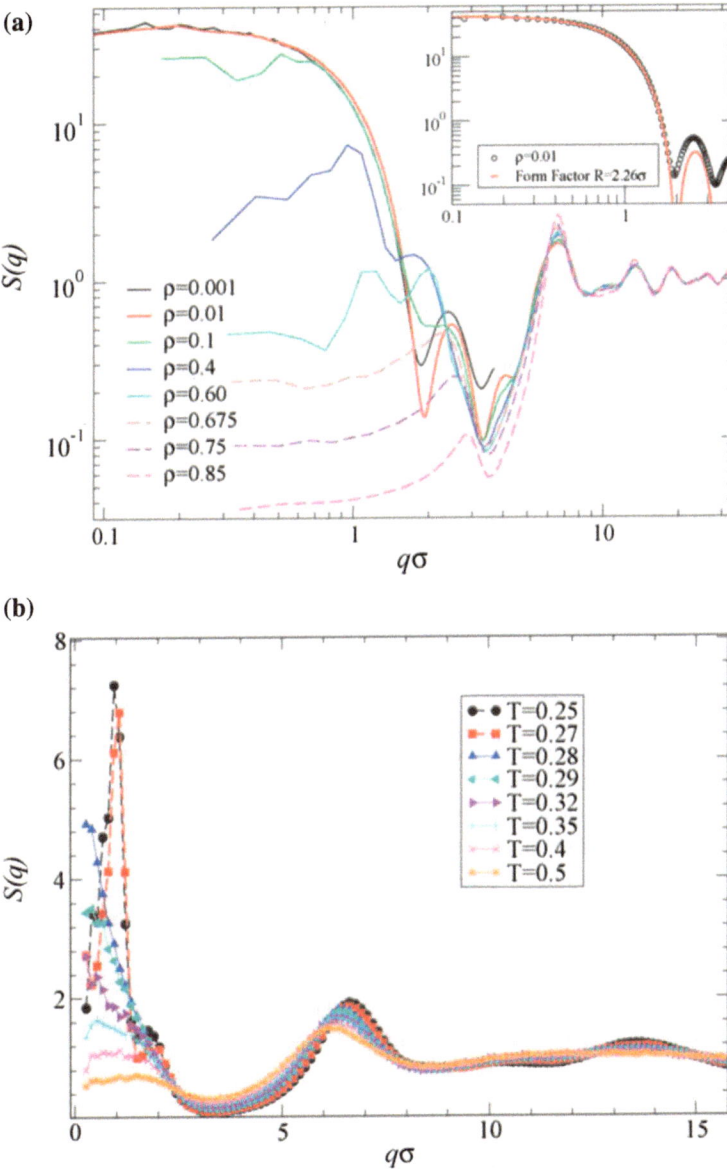

Fig. 1.5 Structure factor for $\Delta/\sigma = 1/2$ at **a** $k_B T/\epsilon = 0.25$ for several densities and at **b** $\rho\sigma^3 = 0.4$ for several $k_b T/\epsilon$. The inset in (a) shows the fit with the form factor of a sphere of radius R. The best-fit value for the radius is $R = 2.26\sigma$. The figure is from Ref. [24] and their q is our k

vesicles are the dominant clusters. The structure factor indeed evolves from the one characteristic of an ideal gas of spherical vesicles (and indeed the inset shows that S(q) is well represented by the form factor of a sphere of radius R) to the one of interacting spheres, in the micelle-rich gas phase. Indeed, on increasing ρ, oscillations at $k\sigma \approx 1.2$ arise, which correspond to distances comparable to the vesicle size. Beyond density $\rho\sigma^3 = 0.6$ the gas condenses into the liquid phase, where only a very weak pre-peak around $k\sigma \approx 3$ is found.

The second panel of Fig. 1.5 shows the evolution of the structure factor on cooling along the $\rho\sigma^3 = 0.4$ isochore. Compared to simple liquids, one observes a non-negligible scattering at small k, which progressively increases on approaching the phase separation. These are the standard critical fluctuations which are expected to diverge on approaching a spinodal line. While in simple liquids, below the spinodal temperature the system phase separates in a gas coexisting with a liquid phase, here, the peculiar shape of the gas-liquid coexistence line (Fig. 1.3) opens up new stable states, composed by interacting vesicles and $S(k)$ becomes peaked at the vesicle-vesicle distance.

1.10 Perspectives

As perspectives we mention how the study of the Janus particles could shed some light on the process by which the simple Janus geometry of the nucleotide determines the DNA structure. DNA elemental nucleotides, strung together to form a strand, contain one portion that is capable of forming hydrogen bonds with others and another part that is inert. The double helix is an intertwined pair of chains with the hydrogen-bond-forming parts pointed inward and the inert parts pointed outward. Protein too, join functionally different elements. They can be charged on one side and hydrophobic on the other, as are the soap molecules, whose like hydrophobic sides attract. Or they can be positively charged in spots but charged in others, in which case opposites attract. The same Janus principle enables DNA's structure to store and protect information necessary to life and is behind proteins' ability to become essential parts of organisms and to participate in virtually every intracelluler process.

Many of the details of Janus particles are not yet well understood as for example: Will they form different crystal phases? [48] How do they arrange themselves at an oil/water interface? How do they interact with cells and proteins? Is it possible to manufacture Janus particles like DNA molecules to store information or like proteins and enzymes to create lock-and-key structures, with more than just one patch on a single particle? How about Janus particles with different shapes? How about going beyond the simple Janus geometry and creating a triblock structure with patches on the two ends of the particle or at any location on the particle as we wish? Wouldn't it be wonderful to have environmentally responsive Janus particles? Imagine if an assembled nanoparticle loaded with DNA or drug molecules disassembles itself once the nanoparticle reaches inside a cell or a tumor site, in response to local change of pH or the presence of biomarkers. This may help kill cancer or cure diseases. Or imagine if a Janus particle can controllably stabilize and destabilize emulsions. These are only some of the possible challenges which may open up a whole new era of colloidal science research.

Chapter 2
Clustering and Micellization in a Janus Fluid

All truths are easy to understand once they are discovered; the point is to discover them.

[Galileo Galilei (1564–1642)]

Abstract Recent Monte Carlo simulations on the Kern and Frenkel model of a Janus fluid have revealed that in the vapour phase there is the formation of preferred clusters made up of a well-defined number of particles: the micelles and the vesicles. A cluster theory is developed to approximate the exact clustering properties stemming from the simulations. It is shown that the theory is able to reproduce semi-quantitatively the micellisation phenomenon.

Keywords Janus fluid · Janus particles · Kern-Frenkel model · Monte Carlo simulation · Cluster theory · Micelle · Vesicle.

2.1 Introduction

In the statistical mechanics of fluids the liquid state is a particularly fascinating one [1, 49]. A liquid is the state where correlations really play an important role. The pioneering work of Berni J. Alder [50] showed that, because of the absence of attractive forces, the hard-sphere fluid admits only a single fluid phase. In order to find the liquid phase it is sufficient to add an attractive square-well to the pair-potential of the hard-spheres. The resulting hard-sphere square-well fluid admits a bell-shaped gas-liquid coexistence curve [51, 52] with the critical point moving at low temperatures and high densities as the attractive well width diminishes. Recently N. Kern and D. Frenkel [22] studied, through computer experiments, a new fluid model made of hard-spheres with patchy square-well attractions. In its simplest version, the *single patch* case, the model only depends on the *surface coverage* χ of the patch and the attraction range. Between the two extreme cases $\chi = 0$, the hard-sphere model,

R. Fantoni, *The Janus Fluid*, SpringerBriefs in Physics,
DOI: 10.1007/978-3-319-00407-5_2, © The Author(s) 2013

and $\chi = 1$, the hard-sphere square-well model, where the particles pair-potential is isotropic, the particles interaction is directional. The $\chi = 1/2$ model is known as the Janus limit, as the particle, like the roman God, has two faces of different functionalities.

Another important process, which may lead to the manifestation of macroscopic phenomena, in certain fluids, is the clustering or association. In 1956, for example, Leon N. Cooper [53] found that the stable state of the degenerate electron fluid in a metal is one in which particles of opposite spin and opposite momentum form pairs. It was then understood that whereas the electrons in a metal form pairs with relative angular momentum zero, in ^{3}He this would be prevented by the hard core repulsion, and that therefore Cooper pairing had to occur in a state of finite angular momentum. In 1961 A. Lenard [54] proved analytically that a two-component plasma living in one dimension undergoes a transition from the conducting to the insulating state by the formation of neutral dimers made of a positive and a negative charge. A two-component plasma living in two dimensions is only stable at sufficiently high temperatures [55]. But if one adds a hard core to the charges it remains stable even at low temperatures where it undergoes the same transition [56]. The hard core gives rise to anyonic statistics for the quantum fluid living in two dimensions [57]. In three dimensions the two-component plasma with a hard core, the so called restricted-primitive model, also undergoes the clustering transition at low temperature and low densities [58, 59]. An example of a one-component Janus fluid undergoing association is the dipolar hard-sphere fluid. Here a particle can be viewed as the superposition of two uniformly charged spheres: a positive one and a negative one [60]. The restricted primitive model has been carefully studied by M. E. Fisher and Y. Levin [61] within the Debye-Hückel (DH) theory, the Debye-Hückel-Bjerrum (DHBj) theory, and a Debye-Hückel-Bjerrum theory augmented by the dipolar-ionic fluid couplings (DHBjDI). With the DHBj theory they observe a gas-liquid binodal curve with the gas branch bent at high density at low temperature. The usual bell shaped behavior is restored in the DHBjDI theory even though the critical point is now much closer to the Monte Carlo data then the one extracted from the simple DH theory.

In their study of the Kern and Frenkel [22] single patch $\chi = 1/2$ Janus case, F. Sciortino et al. [27] found that the gas branch of the coexistence curve bends at high densities at low temperatures. Below the critical point, the fluid tends to remain in the gas phase for a larger interval of densities respect to the $\chi = 1$ case. This behavior is due to the tendency of particles to associate due to the directional attractive component in the pair-potential and form clusters. At low temperatures, these clusters interact weakly amongst themselves because the particles of which they are composed tend to expose the hard-sphere hemisphere on the outside of the collapsed cluster.

By studying the clustering properties of the gas phase of the Janus fluid, F. Sciortino et al. [27] discovered that below the critical temperature there is a range of temperatures where there is formation of two kinds of preferred clusters: the *micelles* and the *vesicles*. In the former the particles tend to arrange themselves into a spherical shell and in the latter they tend to arrange themselves as two concentric spherical shells.

It is important to confront existing cluster theories with these new findings based on computer experiments. In this chapter the Bjerrum cluster theory for electrolytes, later extended by A. Tani [62] to include trimers, has been employed (preliminary results appeared in Ref. [2, 3]) for the description of the exact equilibrium cluster concentrations found in the computer experiment of F. Sciortino et al. [27]. The theory is extended to clusters of up to 12 particles in an attempt to reproduce the micellization phenomenon observed in the simulations around a reduced temperature of 0.27. A different determination of the intra-cluster configurational partition function has been devised in place of the one used by J. K. Lee [63].

The Kern and Frenkel [22] fluid has been used to describe soft matter [8] biological and non-biological materials like globular proteins in solution [22, 26, 64] and colloidal suspensions [19, 22], or molecular liquids [65]. Recently there has been a tremendous development in the techniques for the synthesis of patchy colloidal particles [7, 66] in the laboratory. These are particles with dimensions of $10 - 10^4$ Å in diameter, which obey to Boltzmann statistics.[1] From the realm of patchy colloidal particles stems the family of Janus particles for their simplicity [5, 67]. It is possible to create Janus particles in the laboratory in large quantities [15] and to study their clustering properties [16, 68].

The *micelles* and the *vesicles* are complex structures observed in the chemistry of surfactant molecules analogous to those which may be found in the physical biology of the cell [69].

The chapter is organized as follows: in Sect. 2.2 the fluid model is described, Sect. 2.3 presents the clustering properties of the fluid found in the Monte Carlo simulations of F. Sciortino et al. [27], the cluster theory is presented and developed in Sects. 2.4 and 2.5, in Sect. 2.6 the numerical results from the proposed approximation to the exact results of F. Sciortino et al. [27] are compared, and Sect. 2.7 is for final remarks.

2.2 The Kern and Frenkel Model

As in the work of F. Sciortino et al. [27] the Kern and Frenkel [22] single patch hard-sphere model of the Janus fluid is used. Two spherical particles attract via a square-well potential only if the line joining the centers of the two spheres intercepts the patch on the surface of one particles and the patch on the surface of the other particle. The pair-potential is separated as follows

---

[1] The quantum effects start playing a role when the de Broglie thermal wavelength $\Lambda = \sqrt{2\pi\hbar^2/(k_B T m)}$ becomes comparable to the particle diameter σ. At room temperature this means that the nano-particles should have a mass of the order of 10^{-26} Kg whereas the micro-particles should have a mass of the order of 10^{-32} Kg.

$$v(\mathbf{r}_1, \mathbf{r}_2) = \phi_{SW}(q_{12}) \Psi(\hat{\mathbf{n}}_1, \hat{\mathbf{n}}_2, \hat{\mathbf{q}}_{12}), \tag{2.1}$$

where

$$\phi_{SW}(q) = \begin{cases} +\infty & q < \sigma \\ -\varepsilon & \sigma < q < \lambda\sigma \\ 0 & \lambda\sigma < q \end{cases} \tag{2.2}$$

and

$$\Psi(1,2) = \Psi(\hat{\mathbf{n}}_1, \hat{\mathbf{n}}_2, \hat{\mathbf{q}}_{12}) = \begin{cases} 1 \ if \ \hat{\mathbf{n}}_1 \cdot \hat{\mathbf{q}}_{12} \geq \cos\theta_0 \ and \ -\hat{\mathbf{n}}_2 \cdot \hat{\mathbf{q}}_{12} \geq \cos\theta_0 \\ 0 \ else \end{cases}$$

$$\tag{2.3}$$

where θ_0 is the angular semi-amplitude of the patch. Here $\hat{\mathbf{n}}_i(\omega_i)$ are versors pointing from the center of sphere i to the center of the attractive patch, with ω_i their solid angles and $\hat{\mathbf{q}}_{12}(\boldsymbol{\Omega})$ is the versor pointing from the center of sphere 1 to the center of sphere 2, with $\boldsymbol{\Omega}$ its solid angle. σ denotes the hard core diameter and $\lambda = 1 + \Delta/\sigma$ with Δ the width of the attractive well.

A particle configuration is determined by its position and its orientation.

σ will be used as the unit of length and ε as the unit of energy. Two particles then attract if they are within the range of the square-well potential and if their attractive surfaces are properly aligned with each other, and repel as hard spheres otherwise. A more convenient writing of Eq. (2.1) would then be

$$v(\mathbf{r}_1, \mathbf{r}_2) = \begin{cases} \phi_{SW}(q_{12}) \ \text{when} \ \Psi(1,2) = 1 \\ \phi_{HS}(q_{12}) \ \text{when} \ \Psi(1,2) = 0 \end{cases} \tag{2.4}$$

with $\phi_{HS}(q)$ being the hard-spheres potential equal to $+\infty$ when $q < \sigma$ and 0 otherwise.

The relative ratio between attractive and total surfaces is the coverage χ. This can be determined as follows

$$\chi = \langle \Psi(\hat{\mathbf{n}}_1, \hat{\mathbf{n}}_2, \hat{\mathbf{q}}_{12}) \rangle_{\omega_1, \omega_2}^{1/2} = \sin^2\left(\frac{\theta_0}{2}\right). \tag{2.5}$$

As in the work of F. Sciortino et al. [27] only the *Janus case* $\chi = 1/2$ will be considered.

2.3 Clustering Properties

The Janus fluid just described will undergo clustering as there is a directional attractive component in the interaction between its particles. Moreover at low temperatures the collapsed clusters are expected to interact weakly with each other. This is responsible for the bending at high density of the low temperature gas branch of the gas-liquid

Fig. 2.1 Exact cluster concentrations of the Janus fluid with $\Delta = \sigma/2$ at a reduced density $\rho\sigma^3 = 0.01$ and various reduced temperatures $k_B T/\varepsilon$, from the Monte Carlo simulation of F. Sciortino et al. [27]

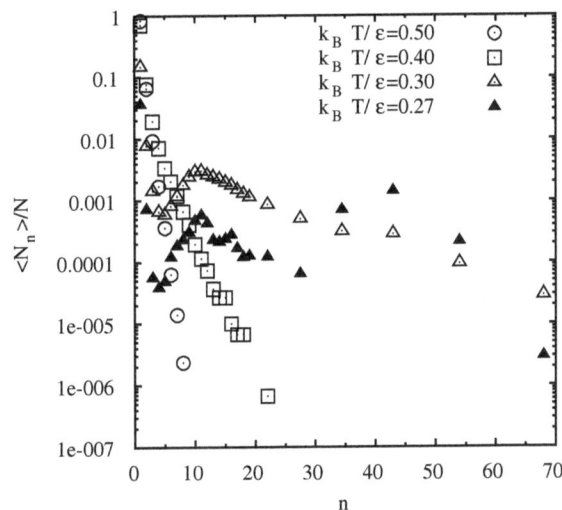

binodal curve recently determined in Ref. [27]. Below the critical temperature, in the vapor phase, the appearance of weakly interacting clusters destabilizes the liquid phase in favor of the gas phase. F. Sciortino et al. during their canonical ensemble (at fixed number of particles N, volume V, and temperature T, with $\rho = N/V$ the density) Monte Carlo simulations of the fluid also studied its clustering properties. In particular they used the following topological definition[2] of a cluster: an ensemble of n particles form a *cluster* when, starting from one particle, is possible to reach all other particles through a path. The path being allowed to move from one particle to another if there is attraction between the two particles. During the simulation of the fluid they counted the number N_n of clusters of n particles, which depends on the particles configurations, and took a statistical average of this number.

Figure 2.1 shows the results they obtained for $\Delta = \sigma/2$ at a reduced density $\rho\sigma^3 = 0.01$ and various reduced temperatures $k_B T/\varepsilon$. From the figure one can see how at a reduced temperature of 0.27, in the vapor phase, there is the formation of two kinds of preferred clusters: one made up of around 10 particles and one made up of around 40 particles.

In their collapsed shape, expected at low temperatures, the particles in the clusters tend to expose their inactive hemisphere on the outside of the cluster, resulting in a weak interaction between pairs of clusters.

In the clusters of around 10 particles the particles tend to arrange themselves into a spherical shell, forming a micellar structure. In the clusters of around 40 particles the particles are arranged into two concentric spherical shells, forming a vesicular structure.

---

[2] Many different ways of defining a cluster may be proposed. This one is the most common one initially used by Gillan in the context of ionic fluids [58, 70].

The aim of the present chapter is to see if one can approximate the exact equilibrium cluster concentrations found in the simulation using a cluster theory. The restriction to clusters made of up to 12 particles is employed to see if the theory is able to reproduce the micellization phenomenon. The theory is described next.

2.4 A Cluster Theory for Janus Particles

Following Ref. [62], the fluid of N particles undergoing clustering as a mixture of N species of clusters is described. Clusters of species $n = 1, \ldots, N$, which will be called n-clusters, are made up of n particles. N_n denotes the number of clusters of species n and with $\rho_n = N_n/V$ their density. The assumption that the chemical potentials of all the cluster species are zero (there is no cost in energy in the formation or destruction of a cluster) is made. Then the grand-canonical partition function of the fluid can be written as

$$Q_{\text{tot}} = \sum_{\{N_n\}}' \prod_{n=1}^{N} \frac{1}{N_n!} \left(q_n^{\text{intra}} \right)^{N_n} Q_{\text{inter}} \left(\{N_n\}, V, T \right), \qquad (2.6)$$

where one separates the coordinates and momenta relative to the center of mass of a cluster from the ones of the center of mass so that q_n^{intra} will be the intra-cluster partition function of the cluster of species n and Q_{inter} the inter-cluster partition function where the clusters are considered as non identical. The prime indicates that the sum is restricted by the condition that the number of particles of the fluid is N,

$$\sum_{n=1}^{N} n N_n = N. \qquad (2.7)$$

Q_{tot} is approximated assuming that the sum can be replaced by its largest dominant contribution. Using the Stirling approximation $N! \approx (N/e)^N$ one then obtains

$$\ln Q_{\text{tot}} \approx \sum_{n=1}^{N} \left[N_n \ln q_n^{\text{intra}} - (N_n \ln N_n - N_n) \right] + \ln Q_{\text{inter}}. \qquad (2.8)$$

The maximum of $\ln Q_{\text{tot}}$ as a function of $\{N_n\}$ on the constraint of Eq. (2.7) is given by the point $\{\overline{N}_n\}$ where the gradients of $\ln Q_{\text{tot}}$ and of the constraint have the same direction. Introducing a Lagrange multiplier λ the equilibrium cluster distribution $\{\overline{N}_n\}$ is then found from the conditions

$$\frac{\partial}{\partial N_n} \ln Q_{\text{tot}} \bigg|_{\{N_n = \overline{N}_n\}} + \ln \lambda^n = 0, \quad n = 1, 2, 3, \ldots \qquad (2.9)$$

The resulting Helmholtz free energy, $\beta F_{\text{tot}} = -\ln Q_{\text{tot}}$, can then be written in terms of the intra-cluster free energy, $\beta f_n^{\text{intra}} = -\ln q_n^{\text{intra}}$, and the inter-cluster partition function as follows

$$\frac{\beta F_{\text{tot}}}{V} = \sum_{n=1}^{N} \left[\bar{\rho}_n \ln \bar{\rho}_n - \bar{\rho}_n \right] + \sum_{n=1}^{N} \bar{\rho}_n \beta f_n^{\text{intra}} + \sum_{n=1}^{N} \bar{\rho}_n \ln V - \frac{1}{V} \ln Q_{\text{inter}}. \quad (2.10)$$

where $\beta = 1/k_B T$ with k_B Boltzmann constant and $\bar{\rho}_n = \overline{N}_n / V$.

The equilibrium cluster concentrations, $\overline{N}_n / N$, are expected to approximate the ones measured in the simulation, $\langle N_n \rangle / N$.

2.5 Relationship Between the Configurational Partition Functions

It will be assumed that Eq. (2.6) also holds at the level of the configurational partition functions Z, as follows

$$Z_{\text{tot}} = \sum_{\{N_n\}}^{\prime} \prod_{n=1}^{N} \frac{1}{N_n!} \left(z_n^{\text{intra}} \right)^{N_n} Z_{\text{inter}} \left(\{N_n\}, V, T \right). \quad (2.11)$$

In the calculation one only works at the level of the configurational partition functions.

Since the clusters are expected to be weakly interacting amongst themselves the inter-clusters configurational partition function will be approximated with: i. the ideal gas (IG) approximation for point wise clusters and ii. the Carnahan-Starling (CS) approximation [71] for clusters of diameter σ_0. A third possibility, not investigated here, would be to use the Boublík, Mansoori, Carnahan, Starling, and Leland (BMCSL) approximation [72, 73] for clusters of different diameters σ_n.

The intra-cluster configurational partition function will be determined through a Monte Carlo (MC) simulation of isolated clusters as described in Sect. 2.5.2. In Fig. 2.2 a flow-diagram of the various methods used to reconstruct the total configurational partition function is shown.

Only a limited number n_c of different cluster species will be used in the analysis. Since one is investigating whether the cluster theory is able to reproduce the micellization phenomenon only the first n_c clusters: $n = 1, 2, 3, \ldots, n_c$ will be considered. Choosing $n_c = 12$.

2.5.1 The Inter-Cluster Configurational Partition Function

Described next are the three approximations used for the calculation of the inter-cluster configurational partition function: i. the ideal gas approximation (IG), ii. the Carnahan-Starling approximation (CS), iii. the Boublík, Mansoori, Carnahan,

Fig. 2.2 Flow-diagram illustrating the methods and the approximations used in reconstructing the total configurational partition function from the inter- and intra-cluster ones

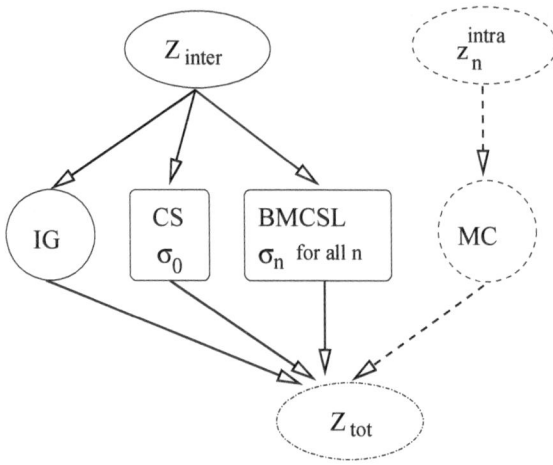

Starling, and Leland approximation (BMCSL). Later on we will only report results which use the IG and CS approximations not the BMCSL.

2.5.1.1 Ideal Gas Approximation

The simplest possibility is to approximate the mixture of clusters as an ideal one so that

$$Z_{\text{inter}} = V^{N_t}, \tag{2.12}$$

where $N_t = \sum_n N_n$ is the total number of clusters.

The equations for the equilibrium numbers of clusters are

$$\overline{N}_n = \lambda^n V z_n^{\text{intra}}, \quad n = 1, 2, 3, \ldots, n_c \tag{2.13}$$

$$N = \sum_n n \overline{N}_n, \tag{2.14}$$

from which one can determine all the concentrations $\overline{N}_n/N$ and the Lagrange multiplier by solving the resulting algebraic equation of order n_c. The case $n_c = 2$ is described in Appendix A.

2.5.1.2 Carnahan-Starling Approximation

A better approximation is found if one uses as the inter-cluster configurational partition function the Carnahan-Starling expression [71] for hard-spheres of diameter σ_0,

$$\ln Z_{\text{inter}} = N_t \ln V - N_t \frac{\eta_t(4 - 3\eta_t)}{(1 - \eta_t)^2}, \tag{2.15}$$

where $\eta_t = (\pi/6)\rho_t\sigma_0^3$ is the clusters packing fraction and $\rho_t = N_t/V$ their density. In this case one needs to solve a system of $n_c + 1$ coupled transcendental equations,

$$\overline{N}_n = \lambda^n V z_n^{\text{intra}} G(\overline{\eta}_t), \quad i = 1, 2, 3, \ldots, n_c \tag{2.16}$$

$$N = \sum_n n \overline{N}_n, \tag{2.17}$$

with $\overline{\eta}_t = (\pi/6)\overline{\rho}_t\sigma_0^3, \overline{\rho}_t = \overline{N}_t/V, \overline{N}_t = \sum_n \overline{N}_n$, and

$$G(x) = \exp\left[-\frac{x(8 - 9x + 3x^2)}{(1 - x)^3}\right]. \tag{2.18}$$

In order to search for the correct root of this system of equations it is important to choose the one that is continuously obtained from the physical solution of the ideal gas approximation as $\sigma_0 \to 0$. Giving a volume to the clusters one introduces correlations between them which will prove to be essential for a qualitative reproduction of the micellization phenomenon though the cluster theory. The Carnahan-Starling approximation amounts to choosing for the sequence of virial coefficients of the hard-spheres, a general term which is a particular second order polynomial and to determine the polynomial coefficients that approximate the third virial coefficient by its closest integer [71]. It could be interesting to repeat the calculation using for the inter-cluster partition function the hard-spheres one choosing all but the first virial coefficient equal to zero, to see if that is sufficient to reproduce the micellization phenomena.

Note that in order to study the vesicles one would have to solve a system of around 40 coupled equations.

2.5.1.3 Boublík, Mansoori, Carnahan, and Starling Approximation

An even better approximation to the CS one is expected to be the BMCSL one where one has to assign n_c cluster diameters σ_n with $n = 1, 2, 3, \ldots, n_c$. In this case one defines the quantities

$$y_1 = \sum_{j>i=1}^{n_c} \Delta_{ij}(\sigma_i + \sigma_j)(\sigma_i\sigma_j)^{-1/2},$$

$$y_2 = \sum_{j>i=1}^{n_c} \Delta_{ij} \sum_{k=1}^{n_c} \left(\frac{\xi_k}{\xi}\right) \frac{(\sigma_i\sigma_j)^{1/2}}{\sigma_k},$$

$$y_3 = \left[\sum_{i=1}^{n_c} \left(\frac{\xi_i}{\xi}\right)^{2/3} x_i^{1/3}\right]^3,$$

and

$$\Delta_{ij} = [(\xi_i\xi_j)^{1/2}/\xi][(\sigma_i - \sigma_j)^2/\sigma_i\sigma_j](x_ix_j)^{1/2},$$

$$\xi = \sum_{i=1}^{n_c} \xi_i, \quad \xi_i = \frac{\pi}{6}\pi\rho_t x_i\sigma_i^3, \quad \sum_{i=1}^{n_c} x_i = 1,$$

where $x_i = N_i/N_t$ is the concentration of clusters of species i.

Then the inter-cluster configurational partition function is given by

$$\ln Z_{\text{inter}} = N_t \ln V + N_t \left\{ \frac{3}{2}(1 - y_1 + y_2 + y_3) + (3y_2 + 2y_3)/(1 - \xi) + \right.$$

$$\left. \frac{3}{2}(1 - y_1 - y_2 - y_3/3)/(1 - \xi)^2 + (y_3 - 1)\ln(1 - \xi) \right\}, \qquad (2.19)$$

the compressibility factor by

$$\frac{1}{\rho_t}\frac{\partial Z_{\text{inter}}}{\partial V} = \frac{(1 + \xi + \xi^2) - 3\xi(y_1 + y_2\xi) - \xi^3 y_3}{(1 - \xi)^3}, \qquad (2.20)$$

and for G one finds

$$\ln G(\xi) = (6(1 - y_1 + y_2 + y_3) - \xi(4 - 6y_1 + 18y_2 + 16y_3$$

$$+ \xi(-5 + 9y_1 - 9y_2 - 13t_3 + \xi(1 - 3y_1 + 3y_2 + 5y_3))))/(1 - \xi)^3/2$$

$$- (1 - y_3)\ln(1 - \xi). \qquad (2.21)$$

Described next is how the intra-cluster configurational partition function z_n^{intra} is determined.

2.5.2 The Intra-Cluster Configurational Partition Function

To estimate the intra-cluster configurational partition function we performed Monte Carlo simulations of an isolated topological cluster.

$u_n^{ex} = \langle \sum_{i<j}^n v(\mathbf{r}_i, \mathbf{r}_j) \rangle/(n\varepsilon)$, the reduced excess internal energy per particle of the n-cluster ($u_1^{ex} = 0$ by definition) was determined as a function of the temperature, and then thermodynamic integration was used to determine the intra-cluster configurational partition function.

It was found that the results for $u_n^{ex}(T^\star)$ can be fitted by a Gaussian as follows

$$u_n^{ex}(T^\star) = a_n e^{-b_n T^{\star 2}} + c_n, \qquad (2.22)$$

with $T^\star = k_B T/\varepsilon$ the reduced temperature.

One can then determine $f_n^{ex, \text{ intra}} = \beta F_n^{ex, \text{ intra}}/n$, given the excess free energy of the n-cluster $F_n^{ex, \text{ intra}}$, as follows

$$f_n^{ex, \text{ intra}}(\beta^\star) = \int_0^{\beta^\star} u_n^{ex}(1/x)\, dx$$

$$= c_n \beta^\star + a_n \sqrt{b_n} \left\{ \frac{e^{-b_n/\beta^{\star 2}}}{\sqrt{b_n/\beta^{\star 2}}} + \sqrt{\pi} \left[\text{erf}\left(\sqrt{b_n/\beta^{\star 2}} \right) - 1 \right] \right\},$$

(2.23)

with $\beta^\star = 1/T^\star$. Calling $v_n = n v_0$ the volume of the n-cluster, with $v_0 = \pi \sigma_0^3/6$, then the intra-cluster configurational partition function is given by

$$z_n^{\text{intra}} = v_n^{n-1} \exp(-n f_n^{ex, \text{ intra}})/n! \approx v_0^{n-1} \exp(-n f_n^{ex, \text{ intra}}),$$

(2.24)

with $z_1^{\text{intra}} = 1$.

Only the first 10 clusters with $n = 3, \ldots, 12$ were studied. The dimer being trivial. To this end the simulation started with an initial configuration of two pentagons with particles at their vertexes juxtaposed one above the other. The two pentagons are parallel to the (x, y) plane, have the z axis passing through their centers, and are placed one at $z = +\sigma/2$ and the other at $z = -\sigma/2$. The particles patches all point towards the origin. The clusters with a lower number of particles were formed by simply deleting particles and the clusters with 11 and 12 particles by adding a particle on the z axis just above the upper pentagon and just below the lower one.

Simulations of the isolated cluster at a fictitious reduced density of $\rho \sigma^3 = 0.05$ were performed, which ensured a simulation box big enough that the cluster did not percolate through the periodic boundary conditions. The results for the excess internal energy calculation for the isolated cluster are compared with the results of F. Sciortino et al. [27] for the low density Janus fluid, from which one extracts cluster information by taking all the clusters found with the same number of particles and averaging their properties, as shown in Fig. 2.3.

At high temperatures the limiting value for the excess internal energy per particle of the isolated n-cluster is $-\varepsilon(n - 1)/n$ corresponding to the stretched cluster. At low temperature ($T^\star < 0.15$) the cluster tends to freeze into certain energy minima. Therefore in order to reach the absolute minimum the following smoothing procedure was used. The Kern and Frenkel potential [22] was smoothed by choosing

$$\Psi(\hat{\mathbf{n}}_1, \hat{\mathbf{n}}_2, \hat{\mathbf{r}}_{12}) = \{\tanh[l(\hat{\mathbf{n}}_1 \cdot \hat{\mathbf{r}}_{12} - \cos\theta_0)] + 1\} \times$$
$$\{\tanh[l(-\hat{\mathbf{n}}_2 \cdot \hat{\mathbf{r}}_{12} - \cos\theta_0)] + 1\}/4,$$

(2.25)

and gradually changing the parameter l, during the simulation, starting from $1/2$ and increasing up to values where there is no actual difference between the smoothed

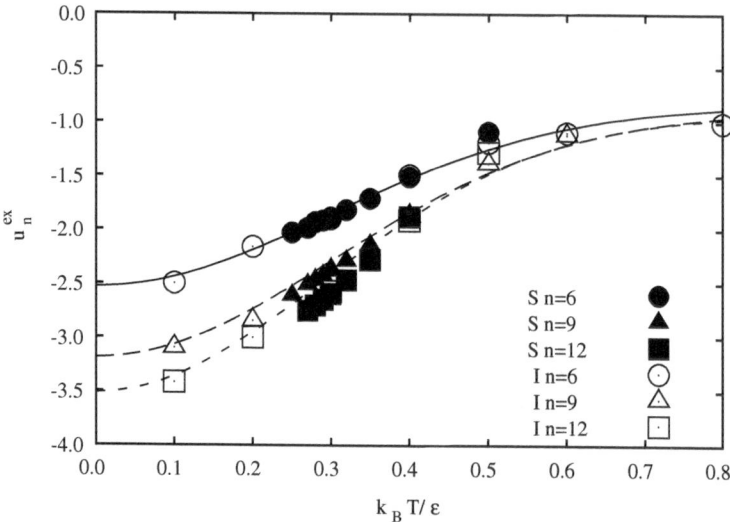

Fig. 2.3 Reduced excess internal energy per particle as a function of temperature for the 6-, 9-, and 12-cluster. The results from the isolated (I) cluster calculation are compared with the results of Sciortino (S) for the Janus fluid with $\Delta = \sigma/2$ at a reduced density $\rho\sigma^3 = 0.01$. The Gaussian fit of Eq. (2.22) is also shown

Table 2.1 The low temperature reduced excess internal energy per particle $\langle V \rangle/(\varepsilon n)$ (V is the potential energy of the cluster) of the clusters with up to 12 particles when $\Delta = \sigma/2$. The gyration radius $R_g^2 = \sum_{i=1}^{n} |\mathbf{r}_i - \mathbf{r}_{cm}|^2/n$ with $\mathbf{r}_{cm} = \sum_{i=1}^{n} \mathbf{r}_i/n$, $\mathbf{r}_i$ being the position of the ith particle in the cluster, is also shown

n	$\langle V \rangle/(\varepsilon n)$	$\langle V \rangle/\varepsilon$	R_g
1	0	0	0
2	−0.5	−1	∼ 1/2
3	−1	−3	∼ 1/$\sqrt{3}$
4	−1.5	−6	0.83
5	−2.0	−10	0.76
6	−2.50	−15	0.75
7	−2.71	−19	0.91
8	−2.88	−23	0.93
9	−3.10	−28	0.96
10	−3.20	−32	1.00
11	−3.36	−37	1.04
12	−3.42	−41	1.08

potential and the original stepwise one. The reduced excess internal energy per particle and gyration radii for such minimum energy configurations are shown in Table 2.1.

In the Metropolis algorithm (see Sect. 1.4.2) used to sample the probability distribution function proportional to $e^{-\beta V}$, where V is the potential energy of the cluster, the random walk moves through the configuration space of the particles forming the cluster through two kinds of moves: a displacement of the particle position and a rotation of the particle (through the Marsaglia algorithm [35]). Two different strategies in the simulations were followed: i. averaging only over the particles configurations that form a cluster and ii. explicitly modifying the acceptance probability by rejecting moves that break the cluster. So in the second strategy all the moves are counted in the averages. The two strategies turned out to give the same results, as they should. The second strategy is preferable to simulate the bigger clusters at high temperature and for small well widths because there is no loss of statistics.

In Appendix B the results for the fit of Eq. (2.22) for the reduced excess internal energy of the isolated clusters are presented.

2.5.3 Thermodynamic Quantities

Once the equilibrium cluster distribution $\{\overline{N}_n\}$ has been determined (within the ideal gas or the Carnahan-Starling approximation for the inter-cluster partition function) the configurational partition function Z_{tot} is known. Then the excess free energy is

$$\beta F^{ex} = -\ln\left(\frac{Z_{\text{tot}}}{V^N}\right), \tag{2.26}$$

the reduced internal energy per particle of the fluid is

$$u = \frac{3}{2\beta^\star} + \frac{1}{N}\frac{\partial(\beta F^{ex})}{\partial\beta^\star} = \frac{3}{2\beta^\star} - \sum_n \frac{\overline{N}_n}{N}\frac{\partial\ln z_n^{\text{intra}}}{\partial\beta^\star} = \frac{3}{2\beta^\star} + \sum_n n\frac{\overline{N}_n}{N}u_n^{ex}, \tag{2.27}$$

and its compressibility factor, in the Carnahan-Starling approximation for the inter-cluster configurational partition function, is

$$\frac{\beta P}{\rho} = \frac{1}{\rho}\frac{\partial\ln Z_{\text{tot}}}{\partial V} \approx \frac{1}{\overline{\rho}_t}\frac{\partial\ln Z_{\text{inter}}}{\partial V} = \frac{1+\overline{\eta}_t+\overline{\eta}_t^2-\overline{\eta}_t^3}{(1-\overline{\eta}_t)^3}. \tag{2.28}$$

Here the approximation $N \approx \overline{N}_t$ has been used, which turns out to be reasonable at the chosen value of the cluster diameter, as shown in Fig. 2.5.

Figure 2.7 shows the results for the compressibility factor and the reduced excess internal energy per particle. The reduced excess internal energy is compared with the Monte Carlo data of F. Sciortino et al. (Fig. 1 in Ref. [27]).

2.6 Results

The numerical results from the cluster theory are presented here and compared with the results of F. Sciortino et al. [27] from the simulation of the Janus ($\chi = 1/2$) fluid with $\Delta = \sigma/2$.

Three different attraction ranges: $\Delta = \sigma/2$, $\Delta = \sigma/4$, and $\Delta = 0.15\sigma$ were studied. To the best of our knowledge there are no Monte Carlo results available for the two smaller ranges.

Only the results obtained from the Carnahan-Starling approximation for the inter-cluster partition function were presented as the ideal gas approximation turned out

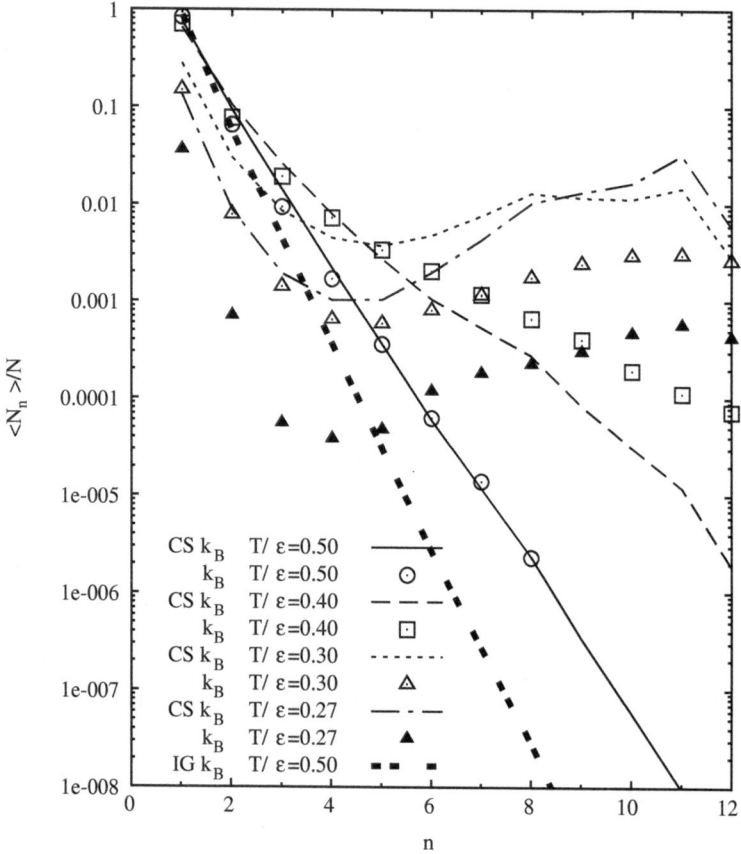

Fig. 2.4 Comparison between the Monte Carlo (MC) data (points) and the Carnahan-Starling (CS) approximation with $\sigma_0 = 2.64\sigma$ (lines) for the cluster concentrations $\langle N_n \rangle / N, n = 1, 2, 3, \ldots, 12,$ as a function of the cluster size n at $\rho\sigma^3 = 0.01$ and various temperatures. The ideal gas (IG) approximation at the same density and the highest temperature $k_B T / \varepsilon = 0.5$ is also shown

to be too crude an approximation even for a qualitative description of the exact clustering properties.

2.6.1 Case $\Delta = \sigma/2$

For $\Delta = \sigma/2$ the following results were found.

2.6.1.1 Equilibrium Cluster Concentrations

In Fig. 2.4 the Monte Carlo data of F. Sciortino et al. [27] (the results reported in Fig. 2.1) and the results from the cluster theory are compared. From the figure one can see that the ideal gas approximation for the inter-cluster partition function is

Fig. 2.5 The compressibility factor, the internal energy per particle, and the logarithm of the total partition function per total number of particles and per total number of clusters as a function of the clusters diameter σ_0 at the thermodynamic state $\rho\sigma^3 = 0.01$ and $k_B T/\varepsilon = 0.5$ for $\Delta = 0.5\sigma$

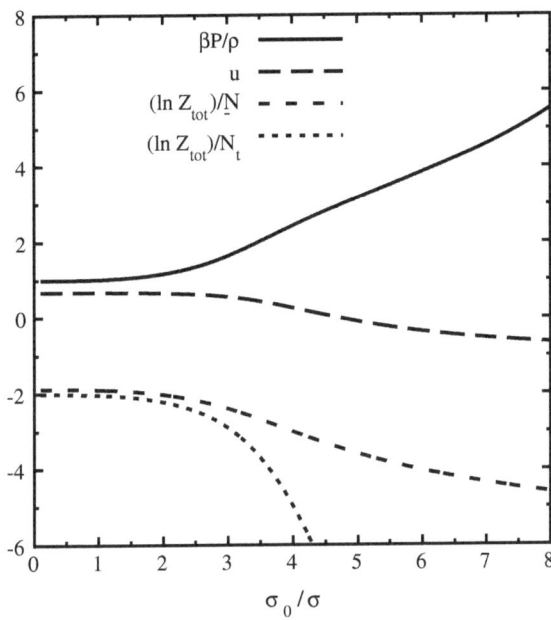

not appropriate even at high temperatures in the single fluid phase above the critical point. In order to find agreement with the Monte Carlo data at high temperatures it is sufficient to give a volume to the clusters, treating them as hard-spheres of a diameter σ_0. In the Carnahan-Starling approximation one gradually increases σ_0 from zero and finds that for $\sigma_0 = 2.64\sigma$ the results of the cluster theory are in good agreement with the Monte Carlo data at $k_B T/\varepsilon = 0.5$. Using the same cluster diameter at all other temperatures, it was seen that the theory is able to qualitatively reproduce the micellization phenomenon observed in the simulation of F. Sciortino et al. [27].

The results also suggest that with a temperature-dependent cluster diameter, or more generally with a cluster diameter dependent on the thermodynamic state of the fluid, better agreement between the used approximation and the exact results could be achieved. The topological definition of a cluster employed has no direct geometrical interpretation. Other definitions with a geometrical nature are possible. For example Lee et al. in their studies of nucleation define an assembly of particles to be a cluster if they all lie within a sphere of radius σ_0 centered on one of the particles. In the simulations of the isolated clusters these have a globular shape at low temperature and a necklace shape at high temperature. The optimal cluster diameter $\sigma_0 = 2.64\sigma$ (found to give good agreement between the exact and approximate clusters concentrations at high temperature) suggests necklace clusters made up of around 3 particles or globular clusters made up of around $2\pi(\sigma_0/\sigma)^2/\sqrt{3} \approx 25$ particles placed on a spherical shell. Since σ_0 is the only free parameter of the theory, it is important to estimate how thermodynamic quantities like the compressibility factor $\beta P/\rho$, the reduced internal energy per particle u, and the logarithm of the total configurational partition function per number of particles, $\ln Z_{tot}/N$, or per number of clusters, $\ln Z_{tot}/\overline{N}_t$, are sensible to variations in σ_0. From Fig. 2.5 one can see

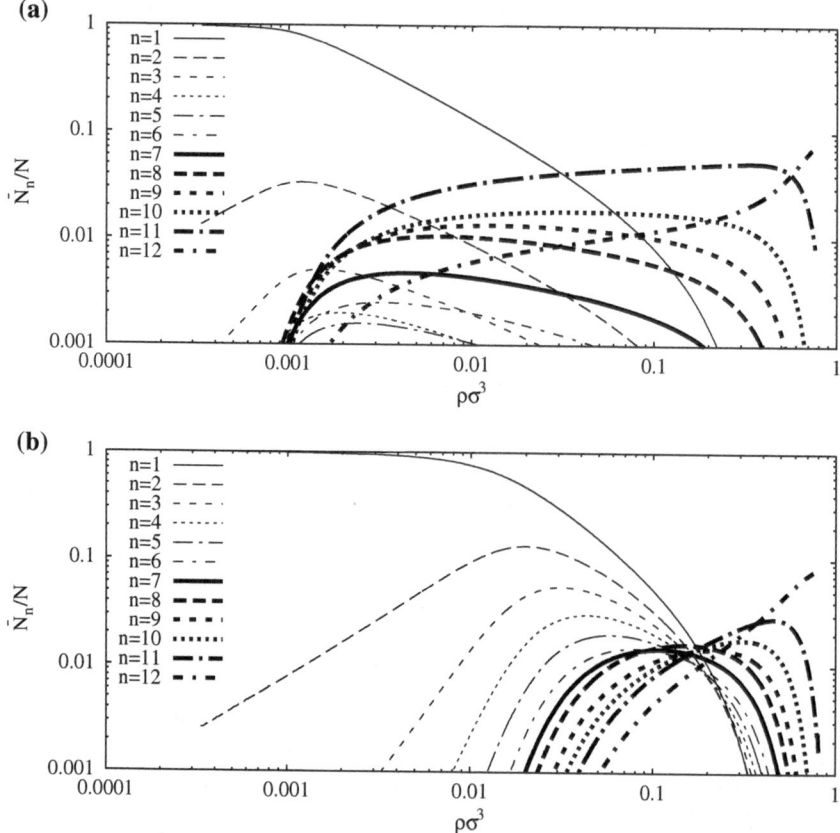

Fig. 2.6 The equilibrium cluster concentrations $\overline{N}_n/N$, $n = 1, 2, 3, \ldots, 12$, as a function of density for $k_B T/\varepsilon = 0.27$ (panel (a)) and $k_B T/\varepsilon = 0.5$ (panel (b)) as obtained from the CS approximation with $\sigma_0 = 2.64\sigma$. Here $\Delta = \sigma/2$

that for the thermodynamic state $\rho\sigma^3 = 0.01$ and $k_B T/\varepsilon = 0.5$, the thermodynamic quantities are roughly independent of σ_0 for $\sigma_0 \lesssim 3\sigma$.

Figure 2.6 shows the behavior of the equilibrium cluster concentrations, from the Carnahan-Starling approximation with $\sigma_0 = 2.64\sigma$, as a function of density at $k_B T/\varepsilon = 0.27$.

From the figure one can see that at very low densities there are essentially no clusters. But as the density increases, clusters of an increasing number of particles appear in the fluid. In particular, at $k_B T/\varepsilon = 0.27$ there is an interval of densities where clusters of 11 particles are preferred.

2.6.1.2 Thermodynamic Quantities

Following Sect. 2.5.3 now the cluster theory within the Carnahan-Starling approximation with $\sigma_0 = 2.64\sigma$ is used to extract thermodynamic information for the Janus

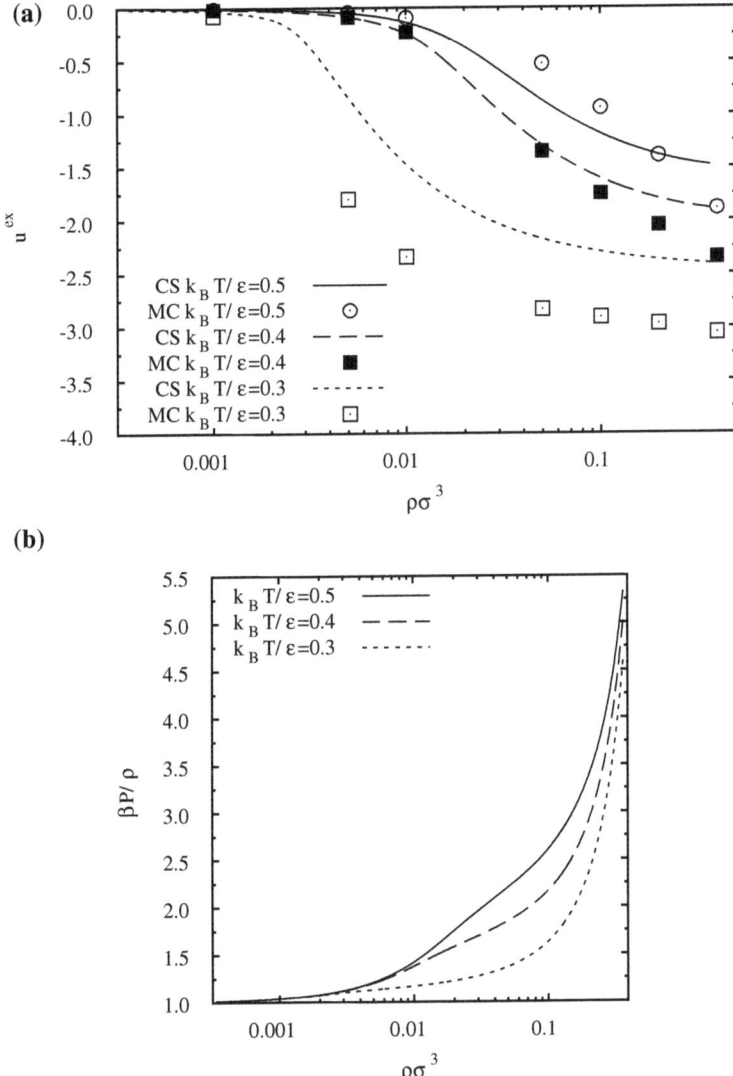

Fig. 2.7 Panel **a** shows the reduced excess internal energy per particle for three different values of temperature as a function of density. The results from the Carnahan-Starling (CS) approximation are compared with the Monte Carlo (MC) results of F. Sciortino et al. [27]. Panel **b** shows the compressibility factor for the same values of temperature as a function of density from the CS approximation (no MC data is available)

fluid. Figure 2.7 shows the results obtained for the excess reduced internal energy per particle and the compressibility factor.

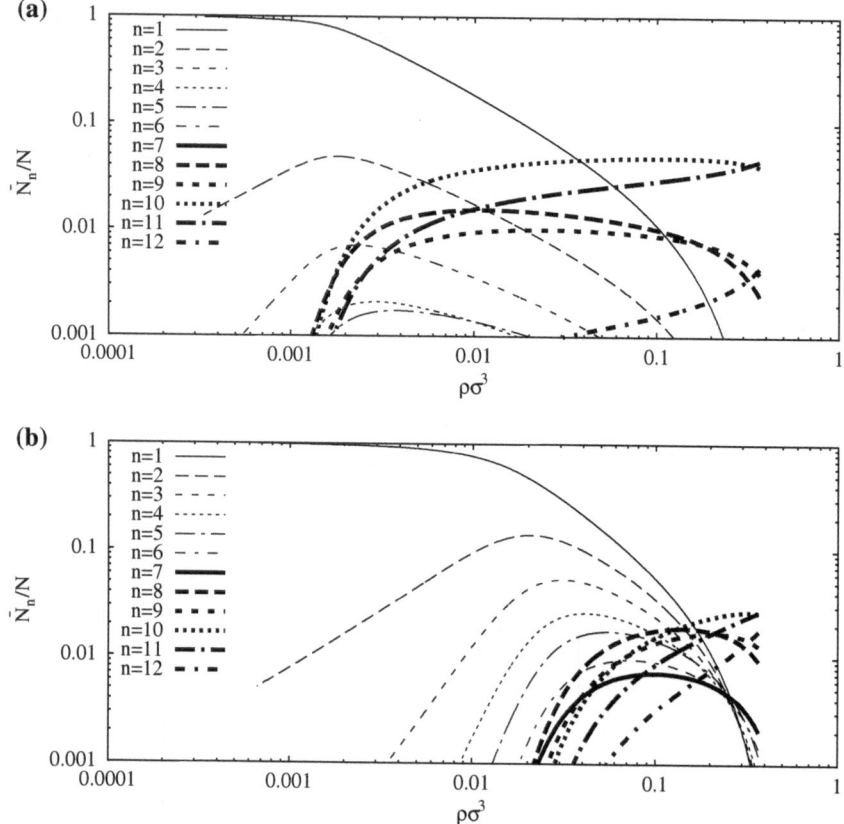

Fig. 2.8 The equilibrium cluster concentrations $\overline{N}_n/N, n = 1, 2, 3, \ldots, 12$, as a function of density for $k_B T/\varepsilon = 0.27$ (panel (a)) and $k_B T/\varepsilon = 0.5$ (panel (b)) as obtained from the CS approximation with $\sigma_0 = 2.64\sigma$. Here $\Delta = \sigma/4$

From Fig. 2.7 one sees that there is a qualitative agreement between the results of the cluster theory and the Monte Carlo results. No Monte Carlo results are available for the compressibility factor.

2.6.2 Case $\Delta = \sigma/4$

Decreasing the width of the attractive well to $\Delta = \sigma/4$ yielded the results shown in Fig. 2.8. One sees that now, at the reduced temperature 0.27, the preferred clusters are the ones made up of 10 particles.

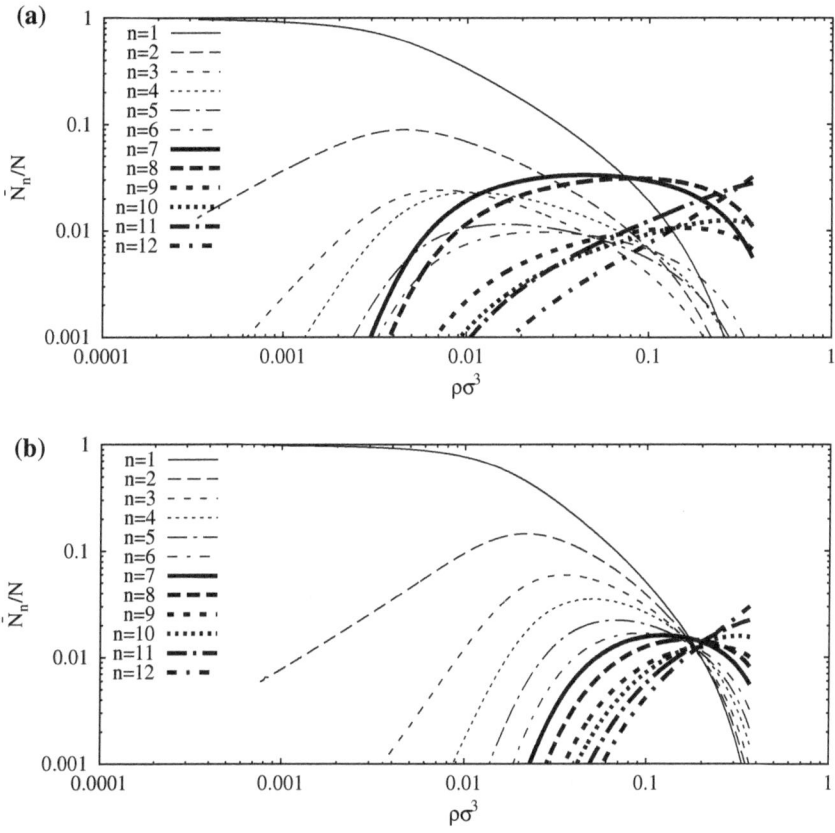

Fig. 2.9 The equilibrium cluster concentrations $\overline{N}_n/N$, $n = 1, 2, 3, \ldots, 12$, as a function of density for $k_B T/\varepsilon = 0.27$ (panel (**a**)) and $k_B T/\varepsilon = 0.5$ (panel (**b**)) as obtained from the CS approximation with $\sigma_0 = 2.64\sigma$. Here $\Delta = 0.15\sigma$

2.6.3 Case $\Delta = 3\sigma/20$

Decreasing the width of the attractive well even further to $\Delta = 0.15\sigma$, the results of Fig. 2.9 were obtained. Now, at the reduced temperature 0.27, there is a range of densities around $\rho\sigma^3 = 0.1$ where the preferred clusters are made up of 7 or 8 particles.

2.7 Conclusions

In this second chapter a cluster theory for a fluid undergoing clustering was constructed and it was shown that it is able to reproduce the micellization phenomena recently observed in the simulation of the vapor phase of Kern and Frenkel Janus particles [27]. A topological definition of the cluster is used [70]. The intra-cluster configurational partition function was determined through thermodynamic integration of the excess internal energy of the cluster, estimated through Monte Carlo simulations of an isolated cluster. In the simulation one restricts the random walk through the configurations of the particles that compose the cluster by rejecting the moves that break the cluster. Due to the geometrical characteristics of the pair-potential it is expected that the clusters, when in their collapsed shape, will be very weakly interacting amongst themselves as the Janus particles will expose the hard-sphere hemisphere on the outside of the cluster. Thus for the estimation of the inter-cluster configurational partition function first the simple ideal gas approximation for point wise clusters was used and then the Carnahan-Starling approximation for clusters seen as hard-spheres of diameter σ_0. The equilibrium cluster concentrations obtained with the ideal gas approximation turned out to disagree, even at high temperatures, with the ones obtained from the simulation of the fluid [27] and were not able to reproduce the micellization phenomenon in the vapor phase. Then σ_0 was gradually increased from zero until good agreement between the equilibrium cluster concentrations obtained with the Carnahan-Starling approximation and the concentrations from the simulation of the fluid [27] at high temperature (above the critical point) was found. Using the same value of σ_0 for lower temperatures (below the critical point) one is able to qualitatively reproduce the micellization phenomenon observed in the simulation of the fluid [27] around a reduced temperature of 0.27 and a reduced density of 0.01. This result is important for two reasons. Firstly it shows that the clustering fundamentally arising from the canonical ensemble description of the fluid of particles can be approximated by a grand canonical ensemble description of a particular clustered fluid. Secondly the second description, which assumes from the start a clustered structure of the fluid, is much less computationally costly than the first. Unlike most previous works on cluster theories where the aim is usually to avoid the Monte Carlo simulation [62, 70], the approach here presented is hybrid and one still uses the Monte Carlo experiment to determine the intra-cluster properties. Of course the goal can only be a qualitative description of the fluid as one specifically prescribe a particular description of the clusters and this is the source of the approximation presented here.

Studying the behavior of the equilibrium cluster concentrations as a function of density and temperature, one observes that the micellization phenomenon only takes place within a particular range of temperatures (below the critical point) and densities (in the vapor phase).

Once the equilibrium concentrations have been found it is possible to determine how the cluster theory approximates the thermodynamic quantities of the fluid. Semi-quantitative agreement between the Monte Carlo data of F. Sciortino et al. [27] and the used approximation for the excess internal energy of the vapor phase was found. For the compressibility factor no Monte Carlo data is available so the results obtained remain a theoretical prediction.

Three different values of the attractive square-well width: $\Delta = \sigma/2$, $\Delta = \sigma/4$, and $\Delta = 3\sigma/20$ were studied. Monte Carlo results [27] are available only for the largest width. This study shows that as the range of the attraction diminishes the micelles tend to be made up of a smaller number of particles.

It would be desirable to study how the analysis changes by implementing the BMCSL in place of the CS to approximate the inter-cluster configurational partition function.

A related interesting problem to that just discussed is the one of trying to give a definition of a liquid drop expected to form in the coexistence region as a result of the condensation instability.

Appendix A
Connection with Wertheim Association Theory

At small χ, allowing only clusters of one (monomers) and two (dimers) particles, one gets

$$\overline{N}_1 = \lambda V z_1^{\text{intra}}, \tag{A.1}$$

$$\overline{N}_2 = \lambda^2 V z_2^{\text{intra}}, \tag{A.2}$$

$$N = \overline{N}_1 + 2\overline{N}_2, \tag{A.3}$$

which is a quadratic equation in λ. The solution for the fraction of patches that are not bonded (fraction of monomers) is

$$\frac{\overline{\rho}_1}{\rho} = \frac{2}{1 + \sqrt{1 + 8\rho\overline{\Delta}}}, \tag{A.4}$$

with $\overline{\Delta} = z_2^{\text{intra}}/[z_1^{\text{intra}}]^2$ and $\rho = N/V$ the density of the fluid, in accord, at low T, with the recent analysis of Sciortino et al. [74] (compare their X of Eq. (10) with our $\overline{\rho}_1/\rho$ and their Δ with our $\overline{\Delta}$), based on Wertheim association theory [75]. The present theory, contrary to the one of Wertheim, allows to consider the case of multiple bonding of the patch.

At high temperature $\overline{\Delta}$ differs from the Δ of Ref. [74] but in this limit the clusters begin to dissociate.

R. Fantoni, *The Janus Fluid*, SpringerBriefs in Physics,
DOI: 10.1007/978-3-319-00407-5, © The Author(s) 2013

Appendix B
The Excess Internal Energy per Particle of the Clusters

Table B.1 gives the fit to the Gaussian of Eq. (2.22) for the reduced excess internal energy per particle as a function of the temperature.

Table B.1 Fit to the Gaussian of Eq. (2.22) for the reduced excess internal energy per particle of the first eleven n-clusters as a function of temperature

	$\Delta = 0.5\sigma$		$\Delta = 0.25\sigma$		$\Delta = 0.15\sigma$		
n	a_n	b_n	a_n	b_n	a_n	b_n	$c_n = -(n-1)/n$
2	0	1	0	1	0	1	−0.5
3	−0.33752	3.88039	−0.3389	6.9050	−0.34559	10.7799	−0.66666
4	−0.77856	4.66976	−0.7706	7.5017	−0.77352	7.97531	−0.75
5	−1.22587	5.16189	−1.0248	5.8901	−1.03428	9.36621	−0.8
6	−1.69844	5.59919	−1.3810	7.3613	−1.20676	9.21365	−0.83333
7	−1.89814	5.26287	−1.4235	6.7666	−1.47964	8.27638	−0.85714
8	−2.06452	5.07916	−1.5201	4.1792	−1.55091	8.50313	−0.875
9	−2.30070	5.47737	−1.5793	4.3672	−1.68144	10.1592	−0.88888
10	−2.39363	5.50909	−1.7253	4.2708	−1.55096	9.41914	−0.9
11	−2.55636	5.64409	−1.8464	4.8294	−1.69591	9.75528	−0.90909
12	−2.59747	6.07744	−1.8541	5.7234	−1.81374	10.5661	−0.91666

R. Fantoni, *The Janus Fluid*, SpringerBriefs in Physics,
DOI: 10.1007/978-3-319-00407-5, © The Author(s) 2013

References

1. Hansen, J.-P., McDonald, I.R.: Theory of Simple Liquids, 2nd edn. Academic Press, New York (1986)
2. Fantoni, R., Giacometti, A., Sciortino, F., Pastore, G.: Soft Matter **7**, 2419 (2011)
3. Fantoni, R.: Eur. Phys. J. B **85**, 108 (2012)
4. Casagrande, C., Veyssie, M.: C. R. Hebdo. Acad. Sci. Paris II **306**, 1423 (1988)
5. Casagrande, C., Fabre, P., Veyssié, M., Raphaël, E.: Europhys. Lett. **9**, 251 (1989)
6. Jiang, S., Chen, Q., Tripathy, M., Luijten, E., Schweizer, K.S., Granick, S.: Adv. Mater. **22**, 1060 (2010)
7. Glotzer, S.C., Solomon, M.J.: Nat. Mater. **6**, 557 (2007)
8. de Gennes, P.-G.: Rev. Mod. Phys. **64**, 645 (1992)
9. Jiang, S., Granick, S.: J. Chem. Phys. **127**, 161102 (2007)
10. Jiang, S., Granick, S. (eds.): Janus Particle Synthesis, Self-assembly and Applications. The Royal Society of Chemistry, Cambridge (2012)
11. Jiang, S., Granick, S.: Langmuir **25**, 8915 (2009)
12. Paunov, V., Cayre, O.: Adv. Mater. **16**, 788 (2004)
13. Yu, H., Chen, M., Rice, P.M., Wang, S.X., White, R.L., Sun, S.H.: Nano Lett. **5**, 379 (2005)
14. Erhardt, R., Boker, A., Zettl, H., Kaya, H., Pyckhout-Hintzen, W., Krausch, G., Abetz, V., Mueller, A.H.E.: Macromolecules **34**, 1069 (2001)
15. Hong, L., Jiang, S., Granick, S.: Langmuir **22**, 9495 (2006)
16. Hong, L., Cacciuto, A., Luijten, E., Granick, S.: Langmuir **24**, 621 (2008)
17. Grnaick, S., Jiang, S., Chen, Q.: Phys. Today **62**, 68 (2009)
18. Gangwal, S.: Directed assembly and manipulation of anisotropic colloidal particles by external fields. Ph.D. thesis, North Carolina State University (2010)
19. Romano, F., Sciortino, F.: Soft Matter **7**, 5799 (2011)
20. Bianchi, E., Blaak, R., Likos, C.N.: Phys. Chem. Chem. Phys. **13**, 6397 (2011)
21. Zhang, Z., Glotzer, S.C.: Nano Lett. **4**, 1407 (2004)
22. Kern, N., Frenkel, D.: J. Chem. Phys. **118**, 9882 (2003)
23. Erdmann, T., Kröger, M., Hess, S.: Phys. Rev. E **67**, 041209 (2003)
24. Sciortino, F., Giacometti, A., Pastore, G.: Phys. Chem. Chem. Phys. **12**, 11869 (2010)
25. Giacometti, A., Lado, F., Largo, J., Pastore, G., Sciortino, F.: J. Chem. Phys. **131**, 174114 (2009)
26. Giacometti, A., Lado, F., Largo, J., Pastore, G., Sciortino, F.: J. Chem. Phys. **132**, 174110 (2010)
27. Sciortino, F., Giacometti, A., Pastore, G.: Phys. Rev. Lett. **103**, 237801 (2009)
28. Reinhardt, A., Williamson, A.J., Doye, J.P.K., Carrete, J., Vareta, L.M., Louis, A.A.: J. Chem. Phys. **134**, 104905 (2011)
29. Fantoni, R., Gazzillo, D., Giacometti, A., Miller, M.A., Pastore, G.: J. Chem. Phys. **127**, 234507 (2007)

R. Fantoni, *The Janus Fluid*, SpringerBriefs in Physics,
DOI: 10.1007/978-3-319-00407-5, © The Author(s) 2013

30. Maestre, M.A.G., Fantoni, R., Giacometti, A., Santos, A.: J. Chem. Phys. **138**, 094904 (2013)
31. Gögelein, C., Romano, F., Sciortino, F., Giacometti, A.: J. Chem. Phys. **136**, 094512 (2012)
32. Shankar, R.: Principles of Quantum Mechanics, 2nd edn. Plenum Press, New York (1994)
33. Fantoni, R., Salari, J.W.O., Klumperman, B.: Phys. Rev. **E85**, 061404 (2012)
34. Hockney, R.W., Eastwood, J.W.: Computer Simulation Using Particles. McGraw-Hill, New York (1981)
35. Allen, M.P., Tildesley, D.J.: Computer Simulation of Liquids. Oxford University Press, Oxford (1987). Appendix G.4
36. Alder, B.J., Wainwright, T.E.: J. Chem. Phys. **31**, 459 (1959)
37. Metropolis, N., Rosenbluth, A.W., Rosenbluth, M.N., Teller, A.H., Teller, E.: J. Chem. Phys. **21**, 1087 (1953)
38. Born, M., Karman, T.V.: Physik. Z. **13**, 297 (1912)
39. Hammersley, J.M., Handscomb, D.C.: Monte Carlo Methods. Chapman and Hall, London (1964)
40. Kalos, M.H., Whitlock, P.A.: Monte Carlo Methods Volume I: Basics, pp. 73–86. Wiley, New York (1986)
41. Frenkel, D., Smit, B.: Understanding Molecular Simulation. Academic Press, San Diego (1996)
42. Panagiotopoulos, A.Z.: Mol. Phys. **61**, 813 (1987)
43. Smit, B., De Smedt, Ph, Frenkel, D.: Mol. Phys. **68**, 931 (1989)
44. Smit, B., Frenkel, D.: Mol. Phys. **68**, 951 (1989)
45. Binder, K.: Z. Phys. **43**, 119 (1981)
46. Wilding, N.B.: Phys. Rev. E **52**, 602 (1995)
47. Panagiotopoulos, A.Z., Quirke, N., Stapleton, M., Tildesley, D.J.: Mol. Phys. **63**, 527 (1988)
48. Vissers, T., Preisler, Z., Smallenburg, F., Dijkstra, M., Sciortino, F.: J. Chem. Phys. **138**, 164505 (2013)
49. Hill, T.L.: Statistical Mechanics. McGraw-Hill, New York (1956)
50. Alder, B.J., Wainwright, T.E.: J. Chem. Phys. **27**, 1208 (1957)
51. Vega, L., de Miguel, E., Rull, L.F., Jackson, G., McLure, I.A.: J. Chem. Phys. **96**, 2296 (1992)
52. Liu, H., Garde, S., Kumar, S.: J. Chem. Phys. **123**, 174505 (2005)
53. Cooper, L.N.: Phys. Rev. **104**, 1189 (1956)
54. Lenard, A.: J. Math. Phys. **2**, 682 (1961)
55. Hauge, E.H., Hemmer, P.C.: Phys. Norvegica **5**, 209 (1971)
56. Kosterlitz, J.M., Thouless, D.J.: J. Phys. C **6**, 1181 (1973)
57. Lerda, A.: Anyons. Springer, New-York (1992)
58. Caillol, J.-M., Weis, J.-J.: J. Chem. Phys. **102**, 7610 (1995). And references therein
59. Valeriani, C., Camp, P.J., Zwanikken, J.W., van Roij, R., Dijkstra, M.: J. Phys. Condens. Matter **22**, 104122 (2010)
60. Rovigatti, L., Russo, J., Sciortino, F.: Phys. Rev. Lett. **107**, 237801 (2011)
61. Fisher, M.E., Levin, Y.: Phys. Rev. Lett. **71**, 3826 (1993)
62. Tani, A., Henderson, D.: J. Chem. Phys. **79**, 2390 (1983)
63. Lee, J.K., Barker, J.A., Abraham, F.F.: J. Chem. Phys. **58**, 3166 (1973)
64. Romano, F., Sanz, E., Sciortino, F.: J. Chem. Phys. **132**, 184501 (2010)
65. Romano, F., Sanz, E., Sciortino, F.: J. Chem. Phys. **134**, 174502 (2011)
66. Pawar, A.B., Kretzschmar, I.: Macromol. Rapid Commun. **31**, 150 (2010)
67. Walther, A., Müller, A.H.: Soft Matter **4**, 663 (2008)
68. Hong, L., Cacciuto, A., Luijten, E., Granick, S.: Nano Lett. **6**, 2510 (2006)
69. Phillips R., Kondev J., Theriot J.: Physical Biology of the Cell. Gerland Science, Taylor & Francis Group, New York (2008) (Problem 9.4)
70. Gillan, M.J.: Mol. Phys. **49**, 421 (1983)
71. Carnahan, N.F., Starling, K.E.: J. Chem. Phys. **51**, 635 (1969)
72. Boublík, T.: J. Chem. Phys. **53**, 471 (1970)
73. Mansoori, G.A., Carnahan, N.F., Starling, K.E., Leland Jr, T.W.: J. Chem. Phys. **54**, 1523 (1971)
74. Sciortino, F., Bianchi, E., Douglas, J.F., Tartaglia, P.: J. Chem. Phys. **126**, 194903 (2007)
75. Wertheim, M.S.: J. Stat. Phys. **35**, 19, 35 (1984), J. Stat. Phys. **42**, 459, 477 (1986)

Index

A
Acceptance probability, 12
Avogadro's number, 10

B
Binodal, 17
Boublík, Mansoori, Carnahan, and Starling
    approximation, 27, 29
Bridge function, 15

C
Canonical ensemble, 5
Carnahan-Starling approximation, 27, 28
Chemical potential, 6
Classical statistical physics problem, 5
Closure, 15
Cluster concentrations, 25
Cluster species, 27
Cluster theory, 23, 26
Clustering, 3, 17, 22, 24
Coexistence, 13, 17, 20–22, 41
Compressibility factor, 30, 33, 37
Configurational partition function, 5, 27
Critical point, 21

D
Density, 6, 7
Detailed balance, 12
Direct correlation function, 15
DNA, 20
Drugs, 3

E
Entropy, 6
Equation of state, 14

F
Gibbs ensemble Monte Carlo, 13, 17

H
Hamiltonian, 5
Hard-spheres fluid, 4, 21
Helmholtz free energy, 5
Homogeneous fluid, 6, 9

I
Ideal gas, 5
Ideal gas approximation, 27, 28
Importance sampling, 11
Indirect correlation function, 15
Inter-cluster partition function, 26
Internal energy, 6, 14, 33, 37
Intra-cluster partition function, 26
Isothermal compressibility, 15
Isotropic fluid, 6

J
Janus limit, 2
Janus balance, 2
Janus fluid, 1
Janus God, 1
Janus particle, 1

K
Kern-Frenkel model, 4, 16–18, 21, 23

L
Lamella, 3, 17
Liquid optics, 3
Liquid state, 21

R. Fantoni, *The Janus Fluid*, SpringerBriefs in Physics,
DOI: 10.1007/978-3-319-00407-5, © The Author(s) 2013

M
Marsaglia algorithm, 13
Metropolis algorithm, 10, 32
Micelle, 3, 17, 22
Mixture, 26
Molecular dynamics, 10
Monte Carlo, 10, 11, 13, 15, 17, 18, 22, 25, 27,
        30, 33, 34, 37, 40

O
Ornstein-Zernike equation, 15

P
Pair correlation function, 6
Pair-potential, 6
Particle densities, 6
Particle synthesis, 2
Partition function, 5
Patch, 21
Periodic boundary conditions, 10
Phase diagram, 17
Pressure, 6, 7, 14
Probability density, 5
Protein, 20

R
Radial distribution function, 9
Random walk, 12
Re-entrant phase diagram, 4
Re-entrant phase diagram, 17

S
Soap molecules, 3
Square-well fluid, 4, 21
Structure factor, 9, 18
Surface coverage, 21

T
Temperature, 5, 7
Thermal average, 5, 11
Thermal wavelength, 5
Time average, 10
Total correlation function, 6
Transition probability, 12
Triblock, 20

V
Vesicle, 3, 17, 22
Volume, 5